COMPUTER RESOURCE book-ALGEBRA

Thomas A. Dwyer

Margot Critchfield

Robert B. Davis, *Editorial Adviser*

HOUGHTON MIFFLIN COMPANY/BOSTON

ATLANTA • DALLAS • GENEVA, ILL. • HOPEWELL, N. J. • PALO ALTO

ABOUT THE AUTHORS

Thomas A. Dwyer is Professor of Computer Science at the University of Pittsburgh, Pittsburgh, Pennsylvania. Dr. Dwyer has taught at the high school level as well as in college, and is Director of Project Solo and Soloworks, experiments in computing for secondary school students.

Margot Critchfield is Senior Research Assistant at Soloworks. She also served as Art and Technical Editor for the curriculum modules developed by Project Solo.

EDITORIAL ADVISER

Robert B. Davis, formerly of Syracuse University, has assumed the positions of Director of the Curriculum Laboratory, Associate Director for Education of the Computer-Based Education Research Laboratory (PLATO Computer Project), and Acting Principal of the Laboratory School at the University of Illinois in Urbana-Champaign.

Illustrations by **Mark Kelley**

1977 Impression

Printed in the United States of America

Library of Congress Catalog Card Number 73-20656

ISBN: 0-395-17889-4

Here are some questions and answers about this book and about computers, and how they relate to the study of algebra.

Q. What's the purpose of this book?

A. Most people would rather do something with a new idea than just hear about it. The purpose of this book is to provide a collection of interesting things that can be done with algebra when a computer is available.

Q. Why a computer?

A. A computer is powerful in ways that are hard to explain—but easy to appreciate once they are experienced. Computers make possible explorations that a person would never attempt with only paper and pencil.

Q. Is a computer essential for using this book?

A. Yes, but it isn't necessary to own a large one. Some schools buy their own minicomputer; others "time share" a remote computer, communicating with it over regular telephone lines. Students and teachers can control both kinds of computers with typewriter-like devices called *terminals.*

Q. Doesn't all that get pretty complicated?

A. Not really. Think of a computer as a very swift (but not too bright) robot that follows instructions exactly. A set of such instructions is called a *computer program.* This book uses a simple programming language called BASIC which makes writing programs easy.

Q. Is an algebra book needed?

A. Yes, since this resource book is intended as a supplement to a regular course in algebra. A simple quiz at the start of each section lets you check your knowledge of the algebraic concepts that are used in the programs of that section.

Q. How should the use of a computer be combined with the other parts of an algebra course?

A. One good way is for each student to work with his or her teacher to select programs as course projects. Many students will also want to devise programs that extend the ideas given in this book. The most important thing is to keep an open mind about what can be done. Computers together with algebra help one build a sort of "mathland," a place where imaginations can poke and prod to see how things work.

Q. Can one start learning how to program immediately?

A. Yes. The Prologue section (pages 1–25) describes the essentials of BASIC, and gives some simple programs you can try right away. Going through the Prologue will also help in the study of algebra. Programming gives experience with concepts like "variable" and "negative number" in a natural setting.

The following table of contents shows how these ideas fit together, and lists the computer programs from which you may choose.

CONTENTS

COACHING AND PRACTICE PROGRAMS

APPLICATION PROGRAMS AND SIMULATIONS

prologue a whirlwind tour of computer programming in BASIC

COMPUTERS can't solve problems by themselves. They must first be given an exact set of instructions, called a *program*. To be "understood" by the computer, such instructions must be written in a special *programming language*. The programming language used in this book is BASIC.

English won't work.... but BASIC will.

The purpose of this Prologue is to introduce the important ideas of programming in BASIC* and provide all the background needed for writing the algebra programs in Section 1. Students who already know BASIC can read through these pages quickly to refresh their memories, and then go right on. Students who don't know BASIC, or who haven't used it recently, should take the time to *run* at least one program from each of the eight parts of this Prologue. Each part contains two practice programs. The odd-numbered programs are easier than the even-numbered ones.

Additional programming techniques in BASIC will be introduced as they are needed in later sections of the book. A compact chart at the beginning of each section lists all the features of BASIC to be used in that section.

Logging in; using RUN, LIST, and SCR

A typical session at a computer terminal consists of four parts:

1. Logging in
2. Entering (or creating) a program
3. Running (or executing) the program
4. Logging out

"Logging in" refers to the process of typing some special "words" at the terminal to let the computer know that you are ready to enter a program. We can't tell you exactly how to do this, since the procedure varies on different systems. We'll show an example of "logging in" on the Time Share Corporation (TSC) system of Hanover, N.H. However, your teacher or a more experienced student will have to show you exactly what to do on your system. Just be sure to do it *yourself* under supervision. That way you'll be sure of the procedure.

*A more detailed explanation of BASIC can be found in the book *A Guided Tour of Computer Programming in BASIC* by Thomas A. Dwyer and Michael S. Kaufman (Boston: Houghton Mifflin Co., 1973).

After you have logged in correctly, most computer systems will print a response including a word like READY. This means that the system is ready to accept your BASIC program. A BASIC program consists of several *statements*. Each statement tells the computer to do something—it's an instruction. So that there will be no confusion about the *order* in which these instructions are to be carried out, you must type a *line number* at the beginning of each statement.

The computer won't carry out (execute) your instructions until you type a special command:

RUN

Let's look at a very simple example:

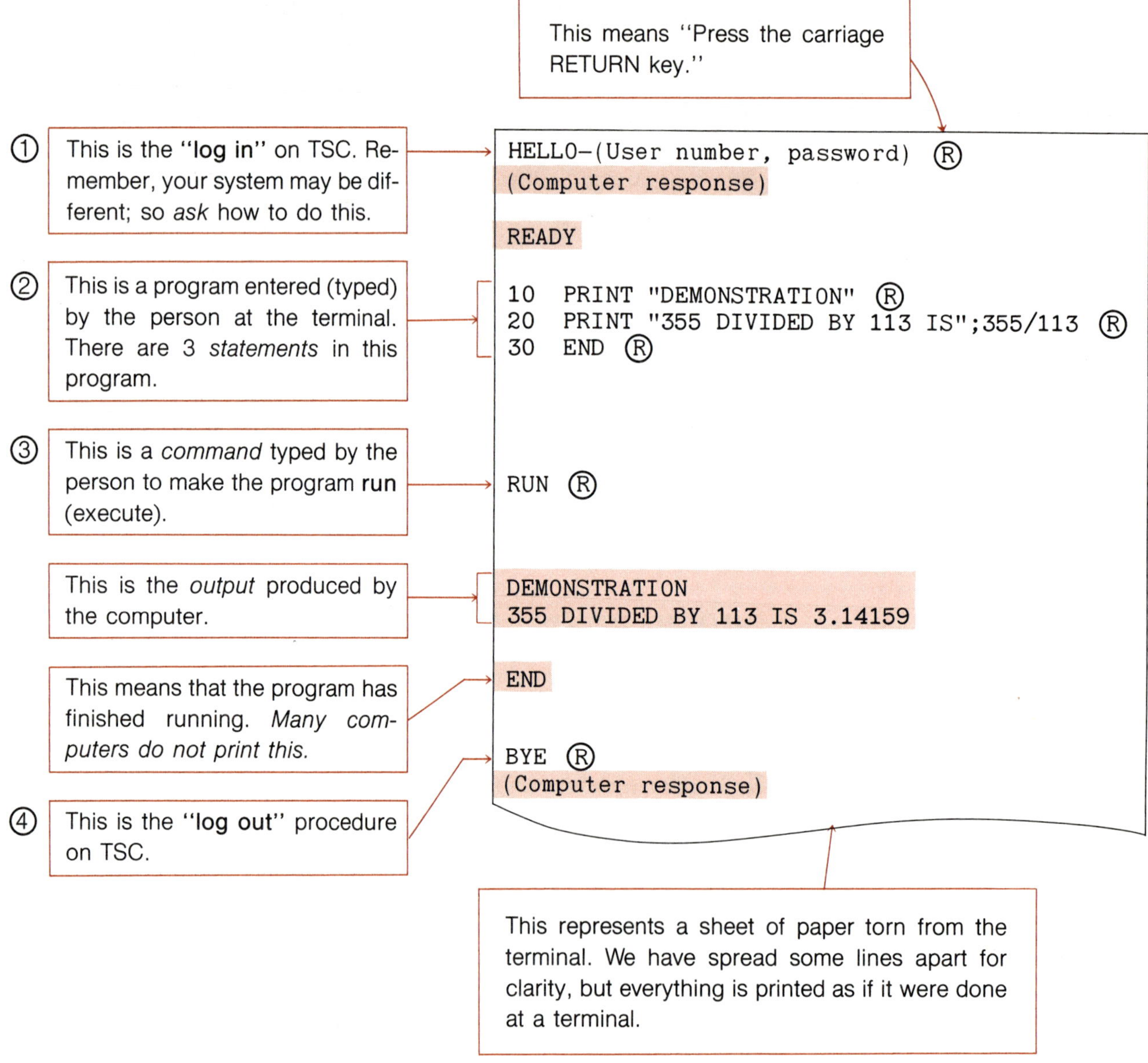

In our example, the portions covered by a screen were typed by the computer. The other parts were typed by the person at the terminal.

Each statement in a program has a line number which is chosen by the programmer. You can use any positive integer from 1 to 9999 for a line number. The computer will always execute the statement with the smallest line number first, the statement with the second smallest line number next, and so on.

If you make a typing mistake, press the carriage RETURN key. You'll probably get a message* telling you that the computer found an ERROR in the line you just typed (which of course you knew). Press the carriage RETURN again, and *type the whole line over again* correctly, as shown in the following example:

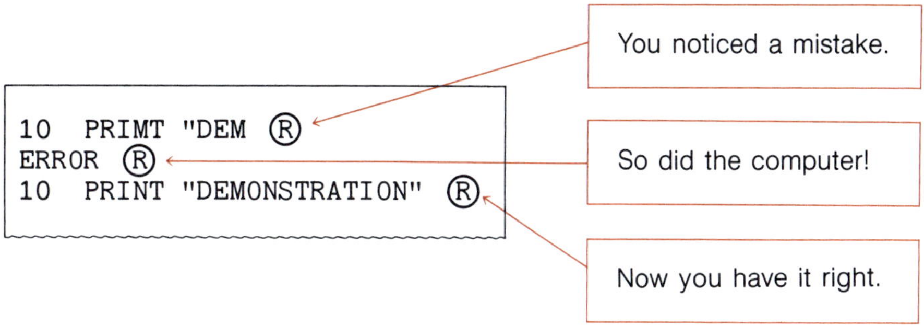

The *second* line 10 replaces the first one. If you'd like to check this, type the command:

LIST

This will show you exactly what statements are in the computer:

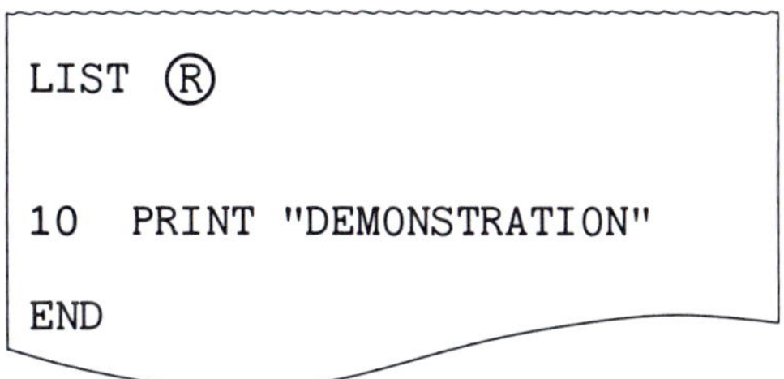

Notice that the first (incorrect) line 10 is now gone. Also notice that commands like RUN and LIST *don't* have line numbers—they're not part of your program.

Special Note: Some computers have "editing" commands which allow you to correct typing mistakes in other ways. You'll have to read about these in the reference manual for your computer system.

NOTE: To delete (get rid of) a statement, just type the line number followed by Ⓡ.

*The wording of these messages varies on different systems.

Suppose that you wish to RUN another program. Do you have to "log out" and then "log in" again? No. All you have to do is to type the command **SCR**(atch) to get rid of the first program, as shown in the following example:

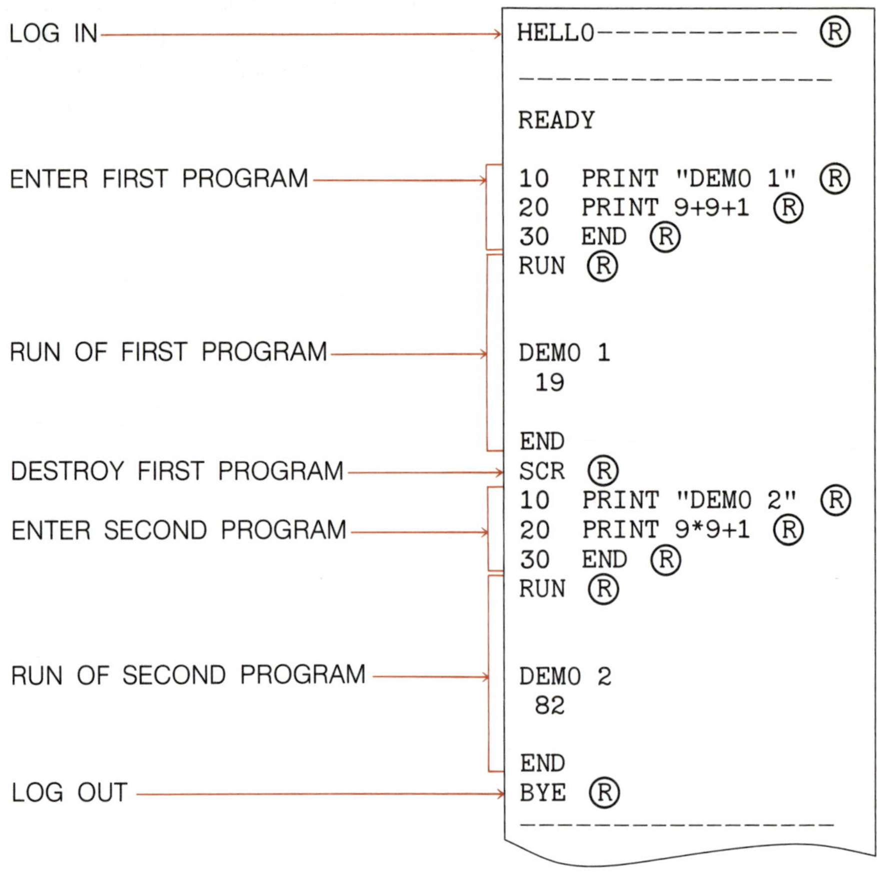

Computer programs use + and − for the operations of addition and subtraction, but they use * and / for multiplication and division. Another operator used by computers is ↑ (called *exponentiation*):

3↑4 is shorthand for 3*3*3*3.

Before going any further in your study of BASIC, we suggest that you try out all these ideas at a computer terminal. Don't be afraid to make mistakes—you won't hurt anything.

We'll give you two programs to try. You don't need to understand how these work yet (that's coming). Just type them in exactly as shown. If you do it right, there will be some interesting output when you type RUN.

Here's what you do:

1. Go to a terminal, and get someone to show you how to turn it on and how to log in.
2. Then type in Program 1 (given below). If you make a typing mistake, press RETURN twice, and then retype the line.
3. Now type RUN Ⓡ.
4. Watch the computer produce its output. If you'd like to see it again, just type RUN Ⓡ again. If you'd like a fresh copy of your program, type LIST Ⓡ.
5. When you're ready to try the next program, type SCR Ⓡ.
6. Now type in Program 2 (given below).
7. Type RUN Ⓡ as often as you wish.
8. When you've finished, log out (on most systems you type BYE Ⓡ, but ask to be sure).

Here are two programs for you to try:

Program 1: BANK (Use this in step 2 above.)

```
10  PRINT "MONEY IN BANK FROM $1000 @ 5% IF INTEREST"
20  PRINT "IS COMPOUNDED ANNUALLY STARTING IN 1970"
30  LET P=1000
40  FOR Y=1 TO 30
50  LET P=P+.05*P
60  PRINT 1970+Y,P
70  NEXT Y
80  END
```

Program 2: BOO-HOO (Use this in step 6 above.)

```
10  FOR K=1 TO 10
20  PRINT "BOO";
30  FOR J=1 TO K
40  PRINT "-HOO";
50  NEXT J
60  PRINT
70  FOR I=1 TO 3+4*K
80  PRINT "-";
90  NEXT I
100  PRINT
110  NEXT K
120  END
```

Using PRINT and END

The statements in BASIC programs must use certain *key words*. In this section we'll look at statements that use the key words PRINT and END.

Statements that use the key word PRINT force the computer to type output at the terminal. This is how you get answers from inside the computer out onto paper. PRINT is also used for printing words, messages, and labels for answers.

Every program in BASIC must have as its last statement an instruction of the form:

```
900   END
```

This tells the computer that there are no further instructions in your program.

> NOTE: From now on we won't show the log-in or log-out procedure—just the program and its RUN. Also we'll not show the Ⓡ which must be typed at the end of every line.

Here's an example:

```
10  PRINT "HOWDY"
20  PRINT "2+2 IS";2+2,"SO LONG"
30  END
RUN


HOWDY
2+2 IS 4        SO LONG

END
```

Notice that anything *inside* quotation marks is printed (typed) exactly as you gave it.

- Notice that when you *don't* use quotation marks, the computer will *first* do the arithmetic called for and *then* print the answer. So if you don't use quotation marks, you must have something that can be printed or calculated as a number.

- Notice that when you want several pieces of output printed on the same line, you use either a semicolon or a comma between them in the PRINT statement:

  (;) The semicolon forces answers to print close together.

  (,) The comma spreads the answers out, putting a new result every 15 spaces.

Here are two programs for you to try. Study them and see if you can predict what will happen when you RUN them. Then try them on a computer.

Program 3: DRILL-SIZES

```
10  PRINT "DECIMAL EQUIVALENTS OF DRILL SIZES"
20  PRINT "1/8 =";1/8
30  PRINT "2/8 =";2/8
40  PRINT "3/8 =";3/8
50  END
```

Program 4: GET-RICH-QUICK

```
10  PRINT "MONEY FROM DOUBLING THE AMOUNT EACH DAY"
20  PRINT "STARTING WITH $1"
30  PRINT "DAY  1=$";1
40  PRINT "DAY  2=$";2
50  PRINT "DAY  3=$";2*2
60  PRINT "DAY  4=$";2↑3
70  PRINT "DAY  5=$";2↑4
80  PRINT "DAY 10=$";2↑9
90  PRINT "DAY 20=$";2↑19
100  PRINT "DAY 30=$";2↑29
110  END
```

Special Note: Computers print large numbers with a compact "scientific notation." When a computer prints 2.14738E+09, it means 2,147,380,000. E+09 means that you are to move the decimal point 9 places to the right. (2,147,380,000 may not be the exact value. For example, the Time Share Corporation system gives such results rounded to 6 significant digits.)

Using LET; variables in BASIC

Suppose that you have just finished a survey of students who used Ultra-Glow toothpaste for the past year. You find that 27 girls and 38 boys got new cavities during the year. Here's how you might tabulate these results on the board: The letter G stands for the number of girls with new cavities, and the letter B stands for the number of boys with new cavities. Next to each letter, the number of students is "stored."

G and B are called *variables* because the numbers we write next to them can *vary*.

We can think of G and B as "labels" for the spaces on the blackboard. These spaces are used to *store* the *values* 27 and 38 in our example.

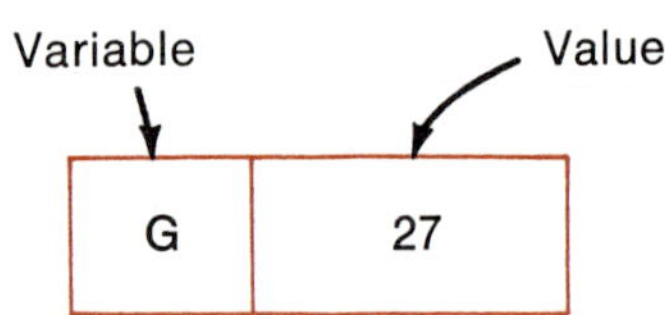

A computer works in a similar way. It has "spaces" inside, which are called *memory locations*. We can give each of these locations a variable name and then store a *value* in the location.

We store values on a blackboard by writing with chalk; we store values in a computer by using a LET statement. Here's how the key word LET works.

The LET statement

```
10  LET G=27
```

does two things. It puts the label G next to one of the computer's memory locations and then stores the number 27 *in* that location. When you want to find out what number is stored in a location, you just say:

> WARNING: This could fool you. It doesn't mean print the letter G—it means print the *number* that is stored in the location *labeled* G.

Here's a program for you to RUN to practice using the LET statement. This program calculates the percent of students with new cavities who were girls, and the percent who were boys.

> *Recall:* If 4 out of 16 people do something, the percent will be (4/16) × 100%, or 25%.

Program 5: CAVITIES

```
10  LET G=27
20  LET B=38
30  LET T=G+B
40  PRINT "TOTAL NO. OF STUDENTS WITH CAVITIES =";T
50  PRINT (G/T)*100;"% WERE GIRLS."
60  PRINT (B/T)*100;"% WERE BOYS."
70  END
```

In BASIC, variables can be single letters of the alphabet—A, B, C, . . . , X, Y, Z—or single letters followed by single digits—A0, A1, A2, . . . , Z7, Z8, Z9.

Now let's look at a program that compares percents of students with new cavities for two brands of toothpaste. Here are the data:

BRAND 1	*BRAND 2*
C1 = 68	C2 = 47
N1 = 102	N2 = 98

C1 is the variable for the number of students having new cavities while using brand 1, C2 is for the number having new cavities while using brand 2, N1 is for the number of students having no new cavities while using brand 1, and N2 is for the number having no new cavities while using brand 2. Here's a program to compare the brands:

Program 6: BRAND-X

```
10  LET C1=68
20  LET N1=102
30  LET C2=47
40  LET N2=98
50  PRINT "% WITH CAVITIES, BRAND 1 =";C1/(C1+N1)*100;"%"
60  PRINT "% WITH CAVITIES, BRAND 2 =";C2/(C2+N2)*100;"%"
70  END
```

Notice that we used parentheses in lines 50 and 60 to make sure that the computer "grouped" the right numbers together. For example,

$$10/(2 + 3) \quad \text{means} \quad 10/5 = 2,$$

but

$$10/2 + 3 \quad \text{would mean} \quad 5 + 3 = 8.$$

On the Time Share Corporation system

$$10/5*2 \quad \text{means} \quad (10/5)*2=4.$$

Check this on your system before using lines 50 and 60 above.

RULE: When there is any doubt about the order in which arithmetic might be done, *use parentheses to make things clear.* The computer will do what's inside the parentheses first.

NOTE: There are usually several ways to write a program. The programs in this book are meant to illustrate ideas, and so they may not always be as compact as possible. If you see a way to improve a program, by all means try it.

D Using INPUT

The trouble with Programs 5 and 6 (pages 8–9) is that they work only for the specific numbers of students we gave. If you wanted to change these numbers, you'd have to change all the LET statements in the program. Is there a better way?

Well, the program would be a lot more flexible if the various numbers of students could be inserted during the RUN of the program rather than during the creation of it. Such numbers are called the *data* for the problem. The INPUT statement allows data to be entered during a RUN. Here's another version of Program 5:

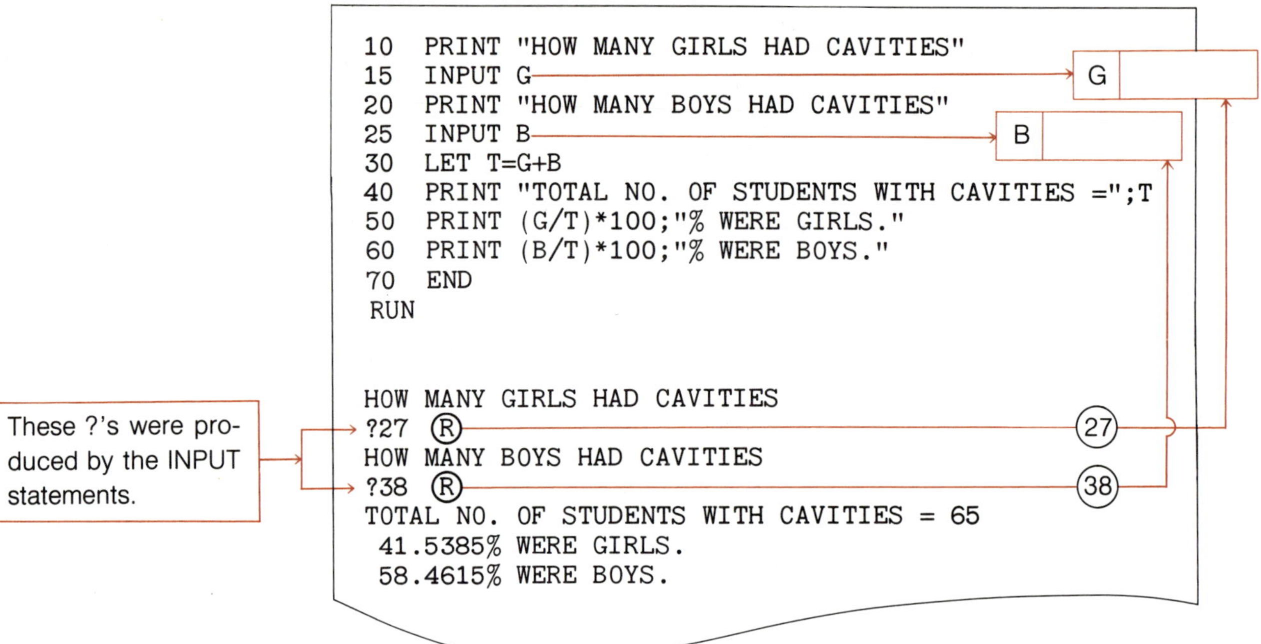

```
10  PRINT "HOW MANY GIRLS HAD CAVITIES"
15  INPUT G
20  PRINT "HOW MANY BOYS HAD CAVITIES"
25  INPUT B
30  LET T=G+B
40  PRINT "TOTAL NO. OF STUDENTS WITH CAVITIES =";T
50  PRINT (G/T)*100;"% WERE GIRLS."
60  PRINT (B/T)*100;"% WERE BOYS."
70  END
RUN

HOW MANY GIRLS HAD CAVITIES
?27 Ⓡ
HOW MANY BOYS HAD CAVITIES
?38 Ⓡ
TOTAL NO. OF STUDENTS WITH CAVITIES = 65
 41.5385% WERE GIRLS.
 58.4615% WERE BOYS.
```


Each INPUT statement causes the computer to print a ? and then *wait* for you to type a number, followed by a Ⓡ. In the program above, 27 was typed after the first ? and the computer stored this 27 under G because the corresponding INPUT statement was:

```
15  INPUT G
```

Similarly, 38 was stored in B.

> IMPORTANT: The numbers you INPUT after each ? must be either integers or numbers containing a decimal. Fractions are not allowed. ?5.75 will be accepted; ?5 3/4 will not be.

> *Special Trick:* Using a semicolon at the end of the PRINT statement preceding the INPUT statement causes the ? to appear on the same line.

Here's an example illustrating this special trick:

```
10  PRINT "HOW OLD ARE YOU";
20  INPUT A
30  PRINT "YOU CAN RETIRE IN";65-A;" YEARS."
40  END
RUN


HOW OLD ARE YOU?14
YOU CAN RETIRE IN 51 YEARS.
```

One INPUT statement can be used to store data in several variables, as shown in the following example:

```
10  PRINT "TYPE YOUR HEIGHT IN FEET, INCHES"
20  INPUT F,I
30  LET H=F*12+I
40  PRINT "YOUR HEIGHT IN INCHES =";H
50  PRINT "YOUR HEIGHT IN CENTIMETERS =";H*2.54
60  END
RUN


TYPE YOUR HEIGHT IN FEET, INCHES
?5,3
YOUR HEIGHT IN INCHES = 63
YOUR HEIGHT IN CENTIMETERS = 160.02
```

It's practice time again. After *predicting* what the following programs will do, RUN them to see if you were right.

Program 7: PHILANTHROPY

```
10  PRINT "HOW MUCH MONEY DO YOU WISH TO GIVE AWAY (IN DOLLARS)";
20  INPUT M
30  PRINT "HOW MANY FRIENDS WOULD YOU EQUALLY DIVIDE IT AMONG";
40  INPUT F
50  PRINT "THEY EACH WOULD RECEIVE";M/F;" DOLLARS."
60  END
```

Program 8: CHICKEN-FEED

```
10  PRINT "HOW MANY QUARTERS, DIMES, AND NICKELS DO YOU HAVE";
20  INPUT Q,D,N
30  PRINT "THAT ADDS UP TO $";.25*Q+.1*D+.05*N;"."
40  END
```

Using IF . . . THEN and STOP

Programs get much more interesting when you use the computer to make decisions. The IF . . . THEN statement in BASIC makes this possible. It tells the computer to do one thing if a certain *condition* is *satisfied,* but to do another thing if that condition is *not satisfied.* The . . . represents the condition.

Here's an example:

```
10  PRINT "WHAT IS YOUR AGE";
20  INPUT A
30  IF A<18 THEN 60
40  PRINT "YOU ARE ELIGIBLE TO VOTE."
50  STOP
60  PRINT "YOU WILL BE ABLE TO VOTE IN";18-A;" YEARS."
70  END
RUN


WHAT IS YOUR AGE?13
YOU WILL BE ABLE TO VOTE IN 5 YEARS.

END
RUN


WHAT IS YOUR AGE?59
YOU ARE ELIGIBLE TO VOTE.
```

Statement 30 is the IF . . . THEN statement. Here's what it means:

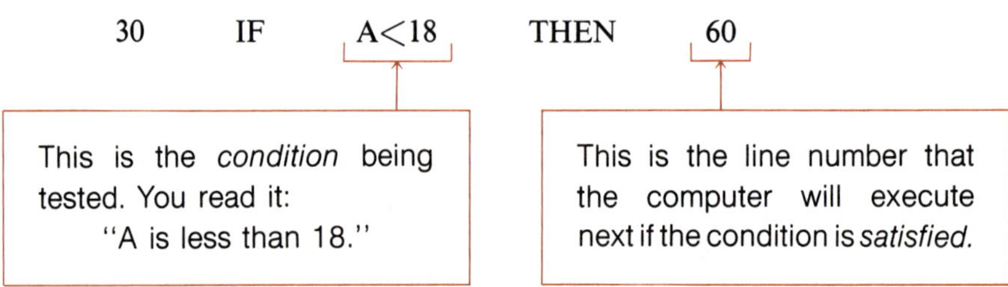

If the condition is *not satisfied* (that is, A is 18, 19, 19.3, . . .), the rule is that the computer will execute the statement right *after* the IF. . . THEN statement. In our example, that is statement 40.

The statement

```
50  STOP
```

means that the computer is to *stop* executing the program at line 50—it should not go on to the END, but stop right where it is. You can have several STOP statements in a program, but only one END, which *must* be the last statement.

Here's a good way to visualize our program in graphical form, using what is called a *flow chart*.

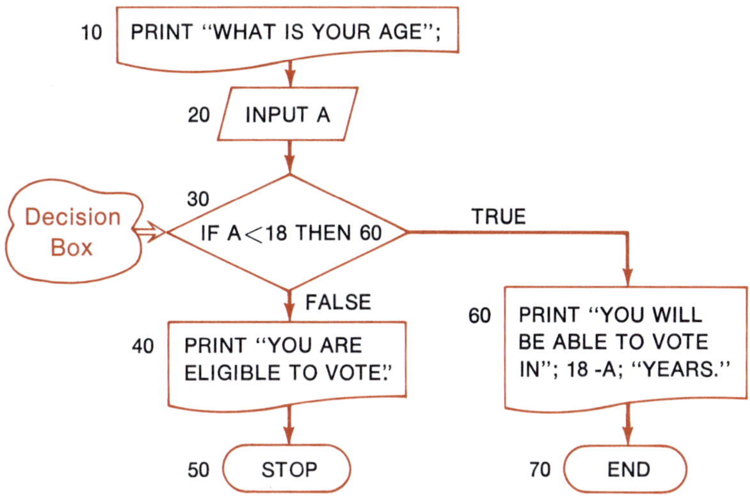

A flow chart uses boxes with different shapes for different purposes. The most important box in our diagram is the diamond-shaped decision box, which shows the two possible *branches* (paths) the computer can take. It represents the IF . . . THEN statement.

Here is how the various conditions are written in BASIC, using the *relations*

$<$, $>$, and $=$:

A < B means "A is less than B."
A > B means "A is greater than B."
A = B means "A is equal to B."

You're also allowed to use the following combinations:

A <= B means "either A is less than B *or* A is equal to B."
A >= B means "either A is greater than B *or* A is equal to B."
A <> B means "A is *not* equal to B."

We won't discuss the less than ($<$), greater than ($>$), or equal ($=$) relations any further here, since they will be covered in algebra. You'll also see later on that a condition is allowed to contain "expressions" as in, for example:

$$(3*X+5)/2 < 5+X$$

Here are two programs for you to study, predict results for, and then RUN. It would be a *very good idea* to draw flow charts for these two programs.

Program 9: ELEVATOR

```
10  PRINT "WHAT FLOOR ARE YOU ON";
20  INPUT F
30  PRINT "TO WHICH FLOOR DO YOU WISH TO GO";
40  INPUT G
50  PRINT "MESSAGE TO ELEVATOR BRAIN:"
60  IF G>F THEN 100
70  IF G<F THEN 120
80  PRINT "THIS CLOWN DOESN'T KNOW WHAT HE WANTS TO DO."
90  STOP
100  PRINT "ENGAGE UP CIRCUITS."
110  STOP
120  PRINT "ENGAGE DOWN CIRCUITS."
130  END
```

Notice steps 60 and 70. There are two conditional statements in succession. If the condition in 60 is false, the program "drops through" to 70. If the condition in 70 is also false, we go to 80. This means that G and F are equal (since if G is not greater than F, and if G is not less than F, then G must equal F).

Program 10: UPPITY-ELEVATOR

```
10  PRINT "THIS ELEVATOR ONLY TAKES PEOPLE UP."
20  PRINT "WHAT FLOOR ARE YOU ON";
30  INPUT F
40  PRINT "TO WHICH FLOOR DO YOU WISH TO GO";
50  INPUT G
60  IF G<F THEN 110
70  IF G=F THEN 140
80  PRINT "THANK YOU. THIS IS YOUR CAPTAIN SPEAKING."
90  PRINT "WE ARE ABOUT TO DEPART--WE WILL GO UP";G-F;" FLOORS."
100  STOP
110  PRINT "YOU HAVE ASKED TO GO ";G-F;" FLOORS."
120  PRINT "THE - MEANS 'DOWN.' THAT'S NOT ALLOWED."
130  STOP
140  PRINT "ARE YOU SURE YOU NEED AN ELEVATOR?"
150  END
```

Using GOTO; interrupting a run

A statement using this key word is easy to understand. For example,

```
60  GOTO 30
```

tells the computer to GO TO (and execute) the statement with line number 30 instead of the normal "next" line.

One use for such a statement is to avoid typing RUN over and over again when we want to use a program repeatedly. Here is an example.

EXAMPLE WITHOUT GOTO:

```
10  PRINT "SQUARE OF A NUMBER N"
20  PRINT "WHAT IS N";
30  INPUT N
40  PRINT N;" SQUARED IS";N*N
50  END
RUN


SQUARE OF A NUMBER N
WHAT IS N?25
 25 SQUARED IS 625

END
RUN


SQUARE OF A NUMBER N
WHAT IS N?13
 13 SQUARED IS 169
END
```

EXAMPLE WITH GOTO:

```
10  PRINT "SQUARE OF A NUMBER N"
20  PRINT "WHAT IS N";
30  INPUT N
40  PRINT N;" SQUARED IS";N*N
45  GOTO 20
50  END
RUN


SQUARE OF A NUMBER N
WHAT IS N?25
 25 SQUARED IS 625
WHAT IS N?13
 13 SQUARED IS 169
WHAT IS N?33
 33 SQUARED IS 1089
WHAT IS N?150
 150 SQUARED IS 22500
WHAT IS N?
```

You can see that the program with GOTO is more convenient, but there is one problem—it will never end! We say that it is caught in an "infinite loop."

All computers have a way of stopping such a program. *The method varies; so you'll have to ask.*

On the Time Share Corporation system, there are two ways to stop a program that is executing:

- If the program has just printed a ? and it is waiting for input, hold down the key marked CTRL (control), and then (still holding CTRL down) press C. This is called "typing a control-C." Follow this by a carriage RETURN.
- At any other time, execution is halted by pressing and releasing the key marked BREAK.

Normally, it is better practice to *program* a way out of the loop. Program 11 below illustrates this technique. After studying the next two programs, try them on your computer.

Program 11: LOOPING

```
10  PRINT "CUBES"
20  PRINT "TYPE A NUMBER GREATER THAN ZERO--EXCEPT"
30  PRINT "TYPE 0(ZERO) WHEN YOU WANT TO STOP."
40  PRINT "WHAT IS YOUR NUMBER";
50  INPUT N
60  IF N=0 THEN 90
70  PRINT N;" CUBED IS";N*N*N
80  GOTO 40
90  PRINT "SO LONG"
100  END
```

Program 12: EXASPERATED

```
10  PRINT "TYPE A 3-DIGIT INTEGER."
20  INPUT N
30  IF N<100 THEN 90
40  IF N>999 THEN 70
50  PRINT "RIGHT!"
60  STOP
70  PRINT "TOO MANY DIGITS--READ THE INSTRUCTIONS!"
80  GOTO 10
90  IF N>-100 THEN 130
100  IF N<-999 THEN 70
110  PRINT "WOW--ARE YOU CLEVER--THAT'S FANTASTIC!"
120  STOP
130  PRINT "NOT ENOUGH DIGITS--READ THE INSTRUCTIONS!"
140  GOTO 10
150  END
```

NOTE: Program 12 provides comments for several wrong INPUTS (lines 70 and 130). But there is one kind of wrong INPUT that is not provided for. If you INPUT a number like 387.5, the program will print "RIGHT." But that isn't right because 387.5 is not an integer. After reading part H and studying Program 16, you should be able to correct this "bug" in the program.

G Using FOR, STEP, and NEXT

The FOR and NEXT statements will undoubtedly become your favorites, since they allow you to get a lot of work done with very little programming effort. These statements are *always* used in pairs as follows:

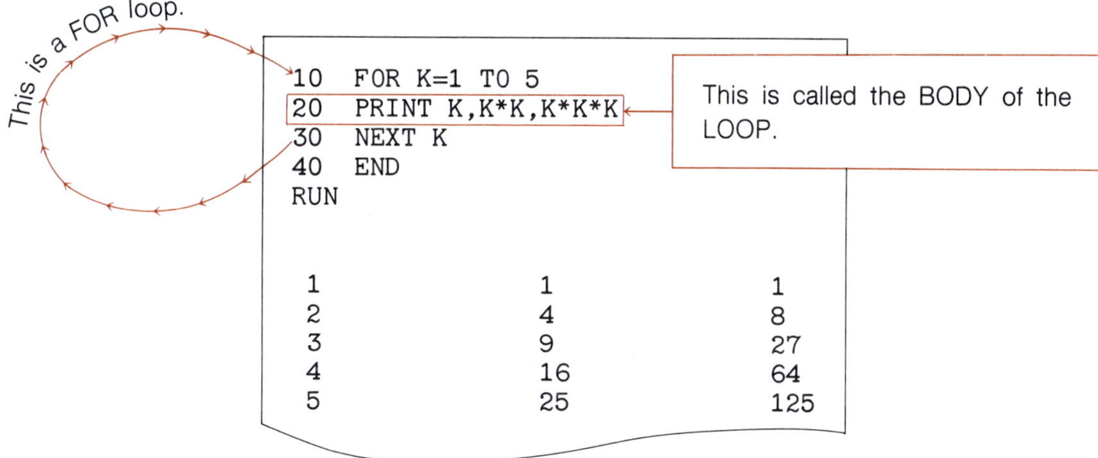


```
10   FOR K=1 TO 5
20   PRINT K,K*K,K*K*K
30   NEXT K
40   END
RUN

 1              1              1
 2              4              8
 3              9              27
 4              16             64
 5              25             125
```

In this program statement 20 is executed five times—for K = 1,2,3,4, and 5. Since the computer keeps going back to line 10 to get the next K, we can say that it is "looping."

If we were to change statement 10 to:

```
10   FOR K=1 TO 5 STEP .5
```

then statement 20 would be executed nine times—

for K = 1,1.5,2,2.5,3,3.5,4,4.5, and 5.

In other words, STEP controls how much is added to K each time the program goes back to line 10. If you don't use STEP, then the computer assumes that the STEP value is 1.

Several BASIC statements can be used in the body of a loop; for example:

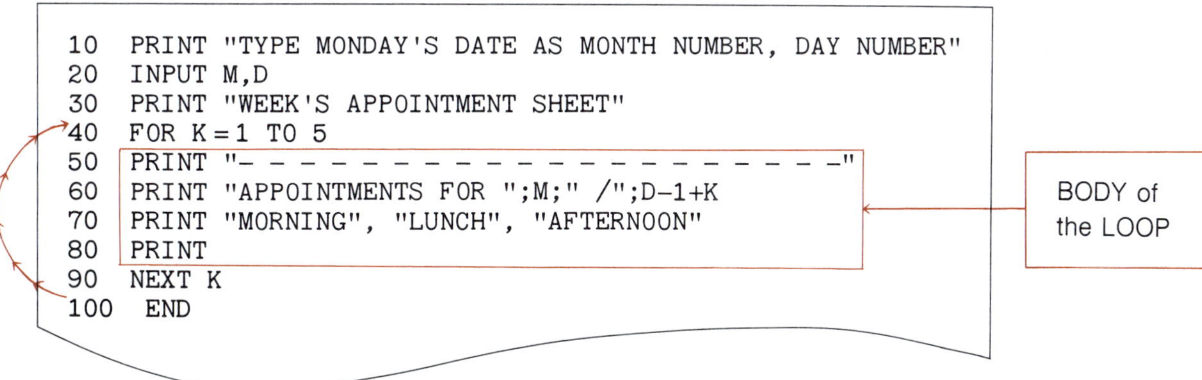


```
10   PRINT "TYPE MONDAY'S DATE AS MONTH NUMBER, DAY NUMBER"
20   INPUT M,D
30   PRINT "WEEK'S APPOINTMENT SHEET"
40   FOR K=1 TO 5
50   PRINT "- - - - - - - - - - - - - - - - - - - - - - -"
60   PRINT "APPOINTMENTS FOR ";M;" /";D-1+K
70   PRINT "MORNING", "LUNCH", "AFTERNOON"
80   PRINT
90   NEXT K
100  END
```

Here's another example to show that several statements can be included in the "body" of the loop:

```
5  PRINT "EXPRESS ELEVATOR STOPS AT"
10  FOR J=10 TO 15
15  IF J=13 THEN 25
20  PRINT J
25  NEXT J
30  END
RUN

EXPRESS ELEVATOR STOPS AT
10
11
12
14
15
```

BODY of the LOOP

The best way to understand the FOR—NEXT pair is to *use* it. Here are two programs for you to study and simulate running (that means pretending you are a computer and writing down the output you would get). Then go to a computer and see what really happens.

Program 13: BLAST-OFF

Version 1

```
10  FOR T=10 TO 0 STEP -1
20  PRINT T
30  NEXT T
40  PRINT "BLAST OFF!!!!!"
50  END
```

Version 2

```
10  FOR T=-5 TO 5
20  IF T=0 THEN 50
30  PRINT "MISSION TIME =";T
40  GOTO 60
50  PRINT "T = 0  ***BLAST OFF***"
60  NEXT T
70  END
```

Program 14: AUTOMATED-DRILL-SIZES

```
10  FOR N=1 TO 16
20  PRINT "     ";
30  PRINT N;"/32 =";N/32,N+16;"/32 =";(N+16)/32
40  NEXT N
50  END
```

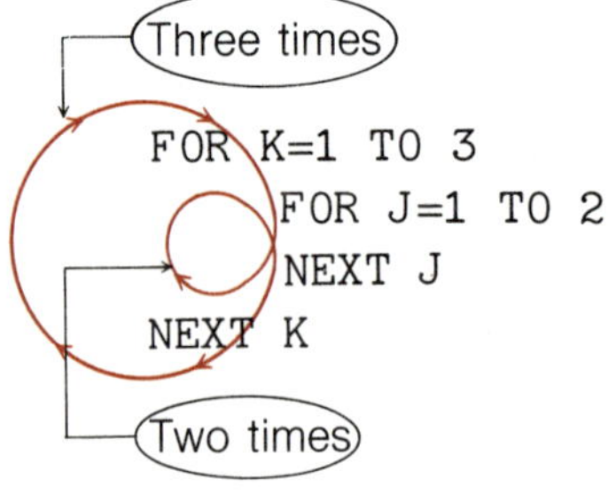

You can put one FOR loop inside another. This is called **nesting.** If you want to, you can then put these **nested FOR loops** inside another FOR loop, and then . . . well, the possibilities boggle the mind!

One way to understand nested loops is to imagine a picture as shown on the left. Once the computer hits the inside loop it must *stay* there, looping around until finished. Then it goes back to the outside loop for its next try. It again hits the inside loop, and must again stay there until finished, and so on.

Here's a simple example showing nested loops:

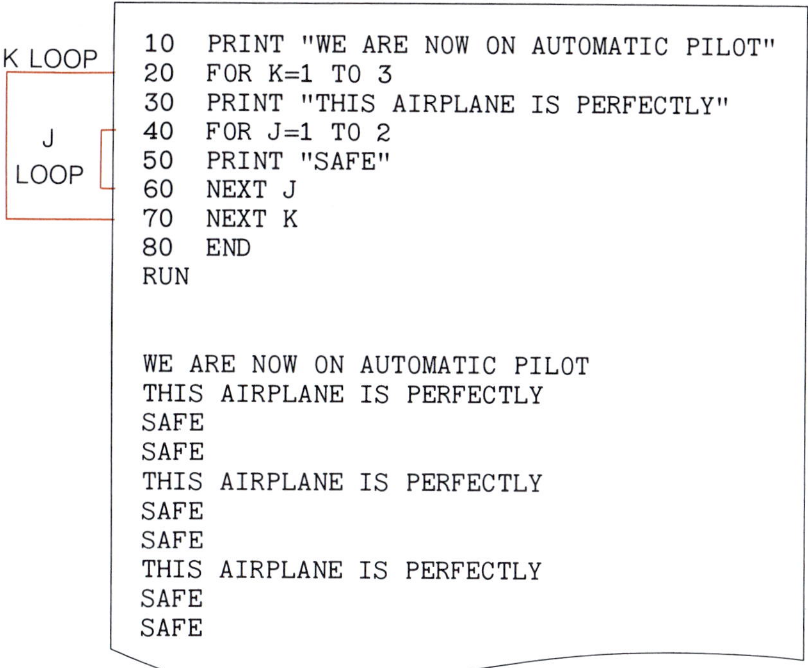


```
10  PRINT "WE ARE NOW ON AUTOMATIC PILOT"
20  FOR K=1 TO 3
30  PRINT "THIS AIRPLANE IS PERFECTLY"
40  FOR J=1 TO 2
50  PRINT "SAFE"
60  NEXT J
70  NEXT K
80  END
RUN


WE ARE NOW ON AUTOMATIC PILOT
THIS AIRPLANE IS PERFECTLY
SAFE
SAFE
THIS AIRPLANE IS PERFECTLY
SAFE
SAFE
THIS AIRPLANE IS PERFECTLY
SAFE
SAFE
```

Here's another example. After you have studied it, you should go back and see if you now understand Program 2.

```
10  FOR K=1 TO 3
20  PRINT "SIGNATURE OF PARTNER NO.";K;":"
30  PRINT
40  PRINT
50  FOR J=1 TO 50
60  PRINT "-";
70  NEXT J
80  PRINT
90  PRINT
100  NEXT K
110  END
RUN


SIGNATURE OF PARTNER NO. 1:


--------------------------------------------------


SIGNATURE OF PARTNER NO. 2:


--------------------------------------------------


SIGNATURE OF PARTNER NO. 3:


--------------------------------------------------
```

This PRINT cancels the effect of the semicolon in line 60. The PRINTS in lines 30, 40, and 90 are used to give blank lines.

Using the SQR, ABS, and INT functions; PRINT TAB

A function can be thought of as a number-processing machine with "IN" and "OUT" ports. BASIC has a number of such functions "built in."

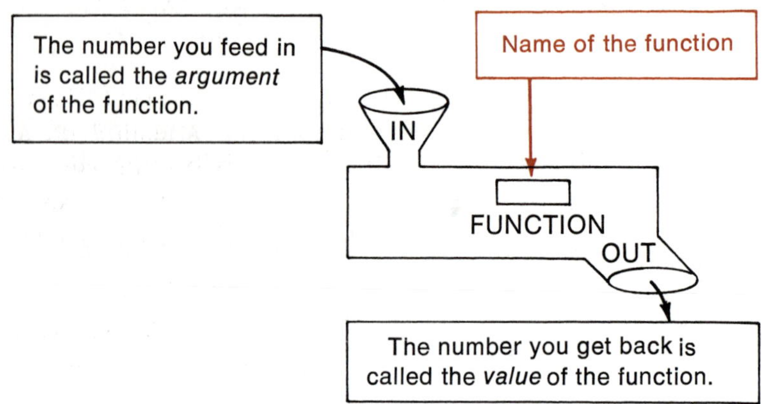

SQR

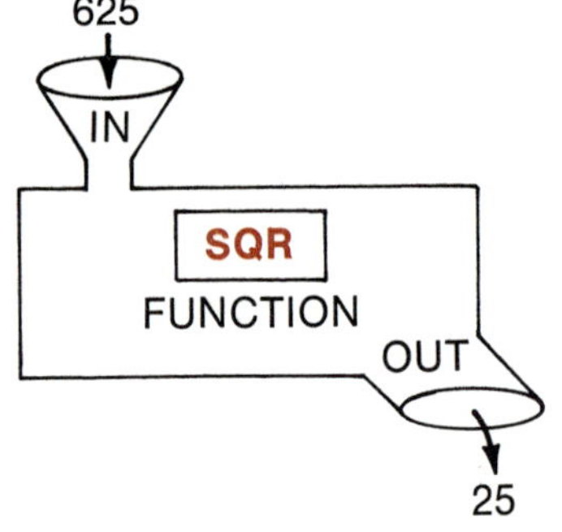

If you feed the SQR (square root) function a positive number, it will give you back the positive square root of that number.

Here is one way to use SQR in BASIC:

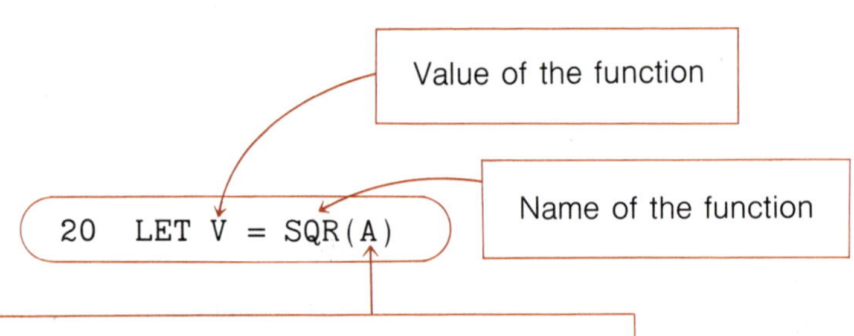

Notice how line 20 is used in the following program:

```
10  INPUT A
20  LET V=SQR(A)
30  PRINT A;" HAS A SQUARE ROOT =";V
40  END
RUN


?625
 625 HAS A SQUARE ROOT = 25
```

If you try to INPUT a negative number, the computer will not accept it. For example:

```
RUN


?-16

SQR OF NEGATIVE ARGUMENT IN LINE 20
```

Functions can be used in LET statements and PRINT statements in the same way that you use a number or a variable, *except*—a function can never be used on the left side of the equal sign in a LET statement.

You can use an arithmetic expression for the argument of a function. The computer does what's inside the parentheses *first*. For example:

```
10  PRINT 10*SQR(3*3+4*4)
20  END
RUN


 50
```

ABS

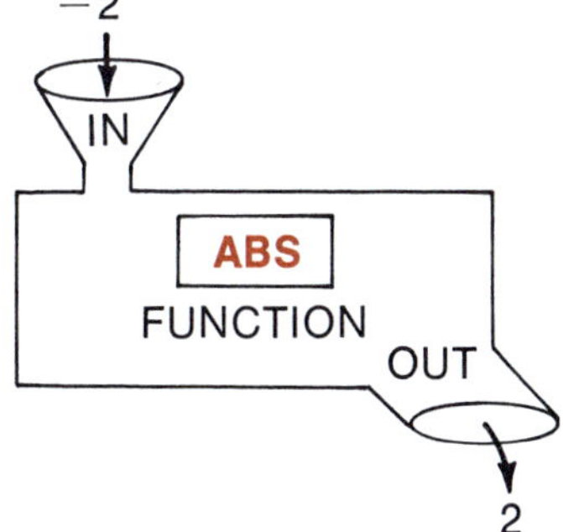

The ABS function gives the *absolute value* of the argument. Study the following example:

```
5  PRINT "X","ABS(X)"
10  FOR N=0 TO 4
20  PRINT 2-N,ABS(2-N)
30  NEXT N
40  END
RUN


X               ABS(X)
 2               2
 1               1
 0               0
-1               1
-2               2
```

INT

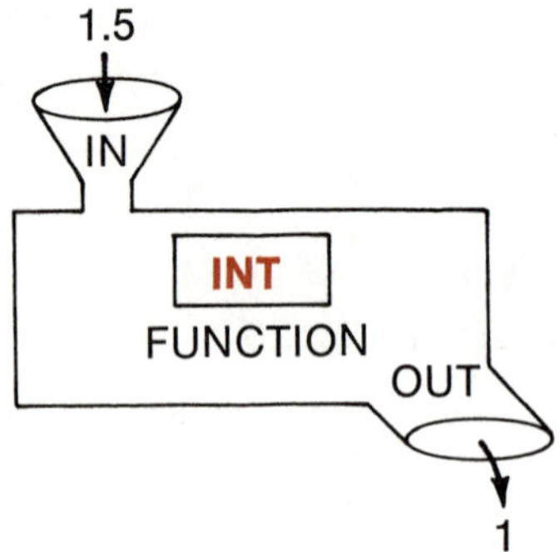

INT gives the "integer part" of the argument. This means that it gives you back the greatest integer less than or equal to the argument. Study the following example:

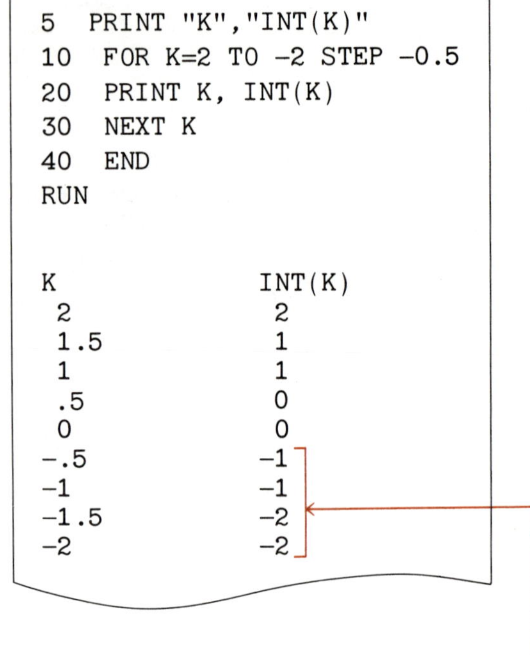

```
5  PRINT "K","INT(K)"
10  FOR K=2 TO -2 STEP -0.5
20  PRINT K, INT(K)
30  NEXT K
40  END
RUN

K              INT(K)
 2              2
 1.5            1
 1              1
 .5             0
 0              0
-.5            -1
-1             -1
-1.5           -2
-2             -2
```

These values may surprise you. Read the explanation of INT carefully to see why they are correct.

One of the uses of INT is to round off decimals to a specified number of places. The following program shows how this works.

```
10  INPUT N
20  PRINT N,INT(N+.5),INT(N*100+.5)/100,INT(N*1000+.5)/1000
30  END
RUN

?3.14159
 3.14159       3              3.14           3.142
```

PRINT TAB

TAB is a feature of BASIC that allows you to produce "fancy" output on the terminal. It's especially valuable in producing *graphs* that make certain mathematical ideas easy to understand. Here's how it works.

The statement

```
10  PRINT TAB(8);"*"
```

tells the computer *what* to print (the symbol *), and *where* to print it (in column 8). To understand the *where*, you must think of the terminal paper as having 72 columns numbered from 0 to 71.

For example:

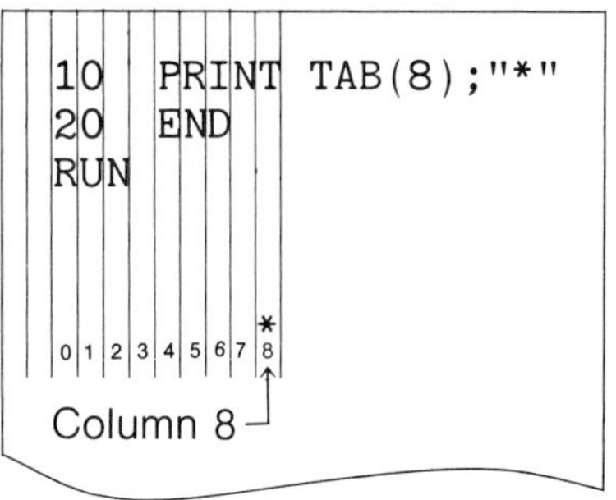

Here's another example:

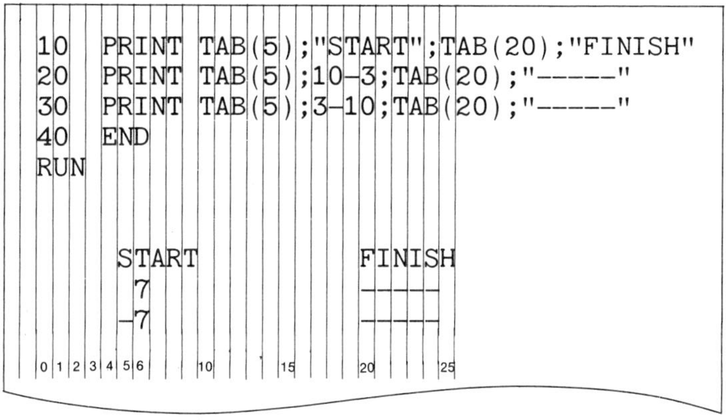

Notice that the computer always leaves a space in front of a *number* for a sign, but it doesn't print + signs. Thus,

PRINT TAB (5);7

puts the 7 in column 6 as shown above.

On the other hand,

PRINT TAB(5);"7"

would put the 7 in column 5 because this 7 is just a symbol.

TAB becomes fun when you use a variable for the column number. For example:

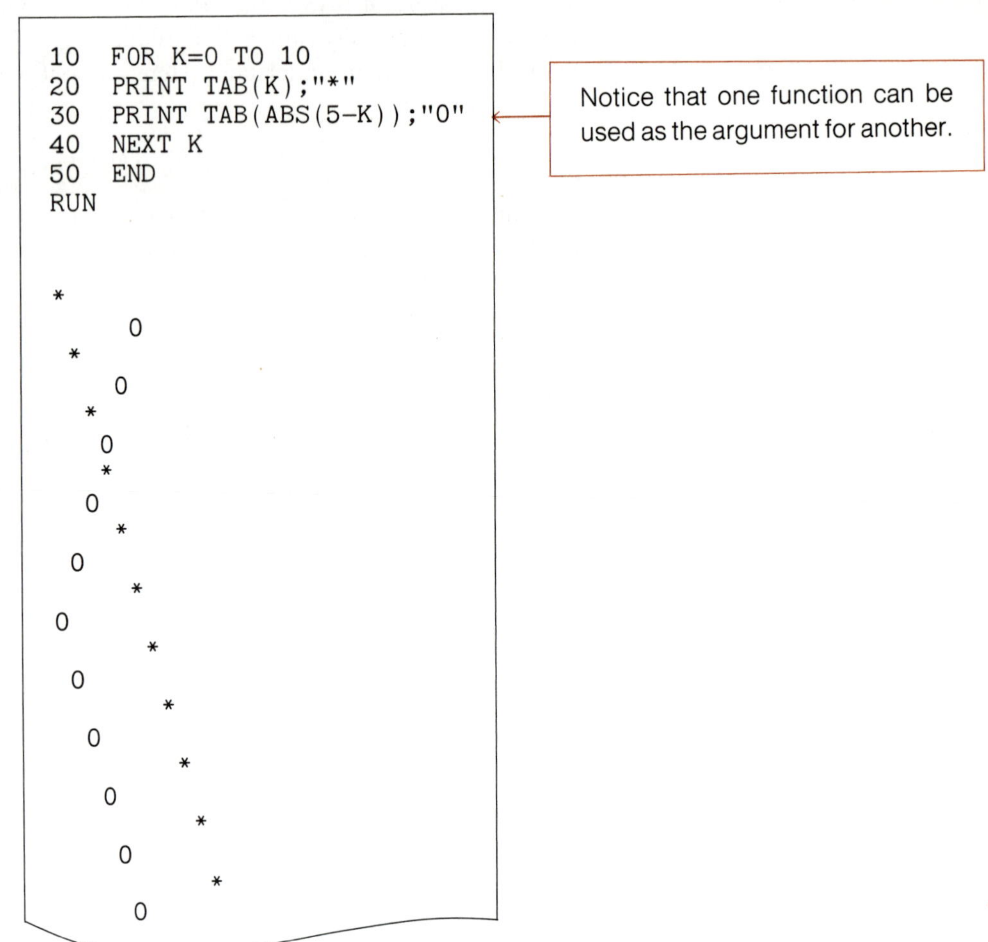

```
10  FOR K=0 TO 10
20  PRINT TAB(K);"*"
30  PRINT TAB(ABS(5-K));"O"
40  NEXT K
50  END
RUN

*
     O
 *
    O
  *
   O
   *
  O
    *
 O
     *
O
      *
 O
       *
  O
        *
   O
         *
    O
          *
     O
```


Study the following programs and then RUN them on your computer.

Program 15: "Z"ANY COMPUTER

```
10  PRINT "ZORRO STRIKES AGAIN"
20  PRINT
30  LET F=0
40  FOR X=0 TO 30
50  PRINT "Z";
60  NEXT X
70  IF F>0 THEN 140
80  PRINT
90  FOR X=30 TO 0 STEP -1
100  PRINT TAB(X);"Z"
110  NEXT X
120  LET F=1
130  GOTO 40
140  END
```

RUN this program and see if you can outsmart it:

Program 16: SWAMI

```
10  FOR K=1 TO 5
20  PRINT "TYPE A POSITIVE INTEGER LESS THAN 1 MILLION:"
30  INPUT N
40  IF ABS(N)>999999. THEN 150
50  IF N=INT(N) THEN 90
60  PRINT "THAT'S NOT AN INTEGER--SHAME ON YOU FOR"
70  PRINT "TRYING TO FOOL A TRUSTING COMPUTER."
80  GOTO 160
90  IF INT(N/2)=N/2 THEN 120
100  PRINT "YOUR INTEGER WAS ODD."
110  GOTO 160
120  PRINT "YOUR INTEGER WAS EVEN--"
130  PRINT "I AM INFALLIBLE."
140  GOTO 160
150  PRINT "READ THE DIRECTIONS!"
160  NEXT K
170  PRINT "FIVE TIMES ARE ENOUGH--SO LONG!"
180  END
```

EXTRA: See if you can use INT to make Program 12 "foolproof."

IT'S ALGEBRA TIME

There are many other useful features of BASIC available, but it will be more interesting to learn them as they are needed. So let's move on to the main part of the book, where we'll combine ideas from algebra and computer programming to do some pretty fancy things.

STEP THIS WAY PLEASE . . .

COMING ★★ ATTRACTIONS

By combining the □ POWER OF THE COMPUTER with a knowledge of ALGEBRA you will be able to

BUILD AN
ULTRAMATIC
ROOT-FINDER in ★ 4 ★ EASY LESSONS

☆ NOT ☆ ONLY ☆ THAT ☆ BUT ☆

You will learn to: ☞ Make & Break Secret Codes, ☞ Write your Own Robot Coaching Programs (guaranteeing yourself ALGEBRAIC SUCCESS! Fame & Fortune), ☞ Build HONEST HAL, the Robot Used-Car Salesman ☞ Unravel Secrets of the ★★★★? FUTURE Using the Punctilious Polynomial Predictor! ☞ Generate Movies & much, much MORE.

SECTION 1
THE LANGUAGES OF ALGEBRA & COMPUTING

Checklist of Computing Skills

We'll summarize the features of BASIC used in each section in chart form. Here's the chart for Section 1:

Previously explained:	PRINT, END, LET, INPUT, IF. . .THEN, GOTO, FOR, NEXT, STEP, INT, PRINT TAB
Requires extra reading:	RND (used in program 1E-4)

Checklist of Algebraic Skills

Each section will start with a short quiz to let you know what knowledge of algebra is needed. Here's the quiz for Section 1:

1. 7*4+3 = __?__
2. 12/12+6*2 = __?__
3. 12/(12+6)*2 = __?__
4. If S=2, then S*S*S = __?__
5. If X=3, then X↑2−3*X = __?__
6. If R=2, then (R+4)*1.5−2*R = __?__
7. Which of the numbers 3, 4.5, −1, −1.5, 0.5, −2.5 belongs in each box?

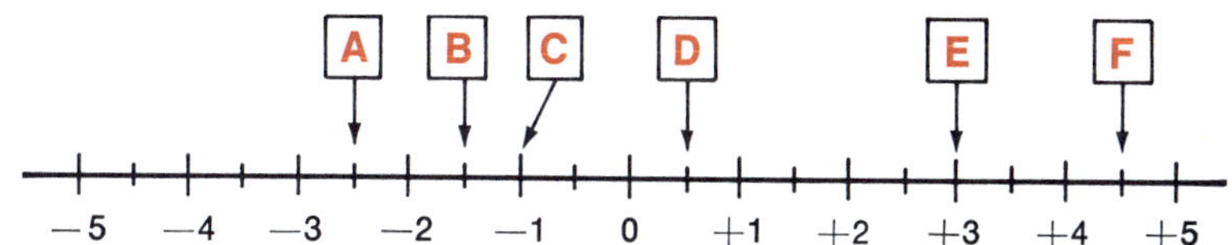

8. Which of the symbols > or = or < belongs in each blank?

   −2 __?__ −2   3 __?__ 1   −4 __?__ −1
   −2 __?__ −5   3 __?__ −1   0 __?__ −1

9. Name all the integers which are > 1 and < 5.
10. Name all the integers which are >= −3 and <= 2.
11. Name all the integers which are > −1 and <= 3.

When you have finished, check your answers with those printed at the bottom of the page, and then ask:

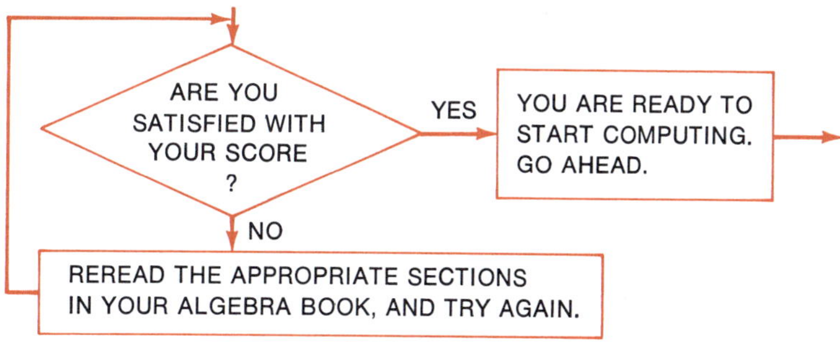

Answers: (1) 31 (2) 13 (3) 4/3 = 1.33333 (4) 8 (5) 0 (6) 5 (7) −2.5 −1.5 −1 0.5 3 4.5 (8) =, >, < >, >, > (9) 2, 3, 4 (10) −3, −2, −1, 0, 1, 2 (11) 0, 1, 2, 3

Unit 1A. Arithmetic Expressions

Building a basketball practice machine like the one shown in our picture would be quite a job. Writing programs that coach a student in algebra is a lot easier.

The best way to prepare for writing a coaching program is to draw a flow chart. Flow charts don't have to use the exact language of BASIC (they're for people, not computers).

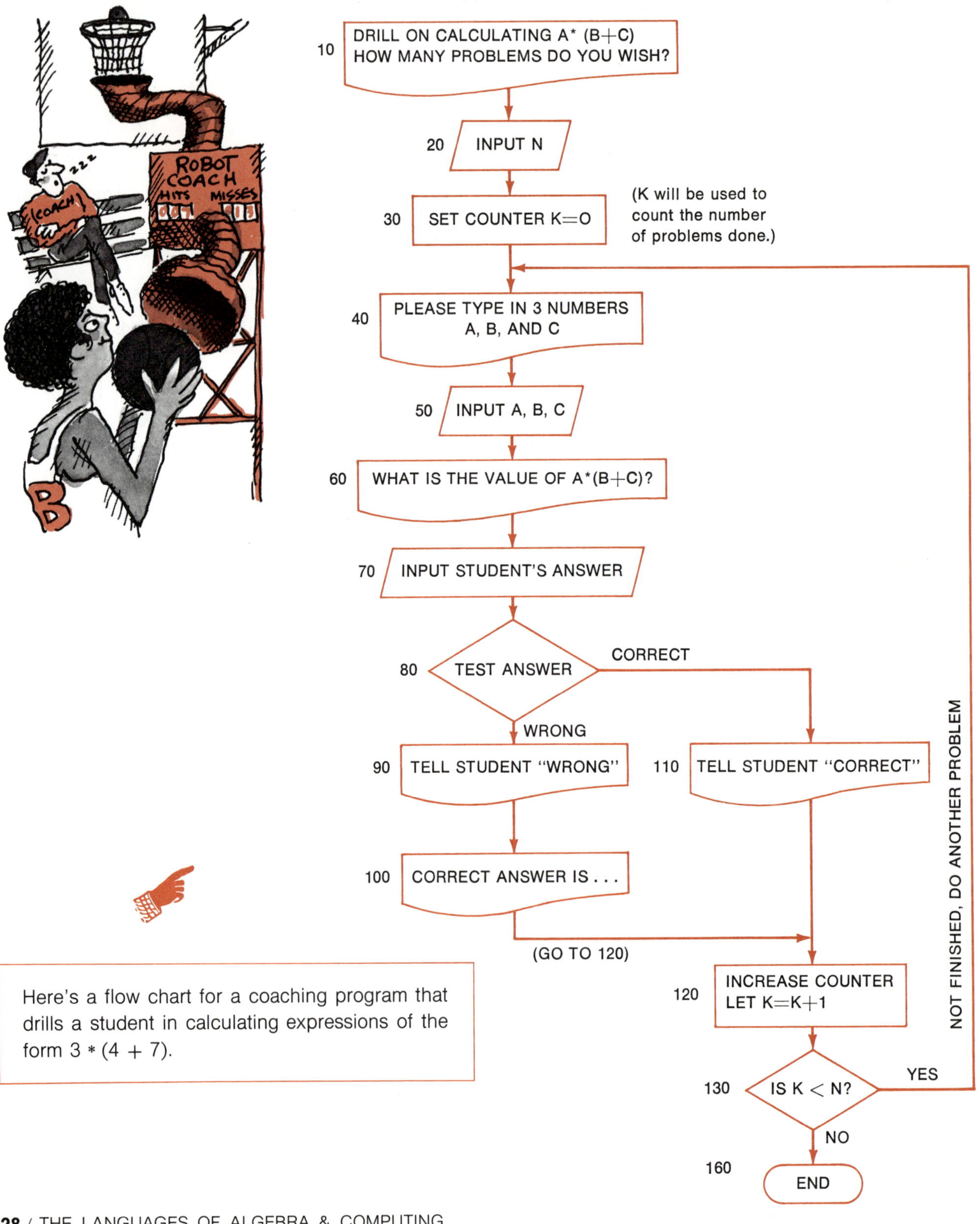

Here's a flow chart for a coaching program that drills a student in calculating expressions of the form 3 * (4 + 7).

Now let's translate our flow chart into a BASIC program, using only those statements that were explained in the prologue section of the book. We'll also give our program a code number and code name. The code number shows the *section, unit,* and *number* of the program within that unit. The code name suggests the content.

1A-1: MULTADD Write a BASIC program that coaches a person in evaluating expressions of the form A * (B + C).

Here's a listing and a sample RUN of a solution to 1A-1:

K is a *counter.* It is set at 0 in line 30, and line 120 increases it by 1 after each problem is done. (Computer programmers call this *incrementing* the counter by 1.)

```
10  PRINT "DRILL ON CALCULATING A*(B+C)"
15  PRINT "HOW MANY PROBLEMS DO YOU WISH";
20  INPUT N
30  LET K=0
35  PRINT
40  PRINT "TYPE IN 3 NUMBERS A, B, C."
45  PRINT "      EXAMPLE: 3,6,2"
50  INPUT A,B,C
60  PRINT "WHAT IS THE VALUE OF";
65  PRINT A;" * (";B;" +";C;") =";
70  INPUT X
80  IF X=A*(B+C) THEN 110
90  PRINT "WHOOPS---THAT'S NOT RIGHT!"
100  PRINT "THE ANSWER IS";A*(B+C)
105  GOTO 120
110  PRINT "CONGRATULATIONS!!! ";
115  PRINT "YOUR ANSWER IS CORRECT!"
120  LET K=K+1
130  IF K<N THEN 35
140  PRINT
150  PRINT "END OF PROGRAM"
990  END
RUN


DRILL ON CALCULATING A*(B+C)
HOW MANY PROBLEMS DO YOU WISH?2

TYPE IN 3 NUMBERS A, B, C.
     EXAMPLE: 3,6,2
?2,5,4
WHAT IS THE VALUE OF 2 * ( 5 + 4) =?18
CONGRATULATIONS!!! YOUR ANSWER IS CORRECT.

TYPE IN 3 NUMBERS A, B, C.
     EXAMPLE: 3,6,2
?7,5,3
WHAT IS THE VALUE OF 7 * ( 5 + 3) =?52
WHOOPS---THAT'S NOT RIGHT!
THE ANSWER IS 56

END OF PROGRAM
```

When K = N, the program ends.

This student asked to do 2 problems. This caused the variable N to have the value 2. The program therefore stopped *after* the counter K became equal to 2.

1A-2: DIVSUB Modify program 1A-1 to practice other computations you have been studying. For example, try drills on expressions like (A + B)/C, A * (B/C), C * (A − B).

1A-3: HELP Modify program 1A-1 so that the student who gets the wrong answer is given further information on how to calculate the correct answer. For example, you might add lines like:

```
101  PRINT "SINCE (";B;" +";C;") =";B+C;" AND"
102  PRINT A;" *";B+C;" =";A*(B+C)
```

Unit 1B. Automatic Scorekeeping

This unit will suggest two ways of making coaching programs more interesting and more helpful.

1B-1: SCORE Modify program 1A-1 so that it automatically keeps score of how well the student has done.

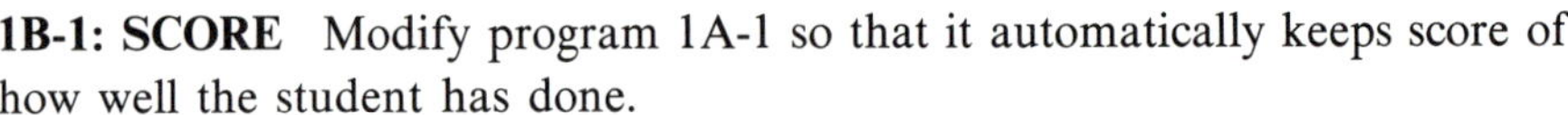

One way to do this is as follows. First add a new statement to program 1A-1:

```
32  LET R=0
```

This sets up a location in the computer called R (we chose R to remind us that it stands for *R*ight answers) and assigns the value zero (0) to R for a start.

Now recall that each time you give a correct answer, program 1A-1 branches to line 110 where the "jackpot" message is printed. So let's squeeze in a new line right after this:

```
118  LET R=R+1
```

This statement causes the computer to increment (increase) the counter R by 1.

Next we have to add some PRINT statements to give the score of right and wrong answers. This can be done by adding the following lines:

```
920  PRINT "YOU DID";N;" PROBLEMS."
930  PRINT "YOU HAD";R;" PROBLEMS RIGHT."
940  PRINT "YOU HAD";N-R;" PROBLEMS WRONG."
950  PRINT "YOUR GRADE IS";(R/N)*100;" PERCENT CORRECT."
960  PRINT "SO LONG,GOOD-BYE,ADIOS,AND AUF WIEDERSEHEN."
```

Note: If program 1A-1 is already in the computer, you need only add lines 32, 118, 920, 930, 940, 950, and 960 before typing RUN. Otherwise you must type in all of program 1A-1 *plus* these new lines.

1B-2: CHEERS Squeeze in a few statements between statements 960 and 990 which print a "cheer-up, try again" message if the percent correct is less than 70% or a congratulation message if the percent correct is > = 70%.

Unit 1C. Evaluating Formulas

1C-1: BLASTO The circus is in town, and El Blasto the magnificent is again thrilling crowds by being shot straight up from the mouth of a cannon.

If he leaves the cannon at a velocity of 14.7 meters per second (which is about 33 miles per hour), his velocity after that (neglecting air resistance) is given by the formula:

$$V = 14.7 - 9.80\ T$$

where T is the time in seconds after "blast-out."

Run the following computer program to find El Blasto's speed for each tenth of a second after "blast-out," for a period of 4 seconds.

```
10  PRINT "T (SEC)","V (M/SEC)","V1 (KM/HR)"
20  FOR T=0 TO 4 STEP .1
30  LET V=14.7-9.80*T
40  PRINT T,V,(V*3600)/1000
50  NEXT T
60  END
```

Some of the values of V will have "–" signs. This means that El Blasto is going *down.*

SCIENTIFIC NOTATION

Back on page 7, we saw that large numbers are printed in the "E" scientific notation. When you run the program BLASTO, you'll find that very small numbers are also printed in scientific notation. In both cases this is done to save space on the terminal paper. Here are two examples to show you how to read scientific notation:

[1] Suppose that the computer prints | 2.37590E+09 |

What **E+09** means is "multiply by 10 with **E**xponent **9**."

Since $10^9 = 1{,}000{,}000{,}000$, 2.3759E+09 means

$$2.3759 \times 1{,}000{,}000{,}000 = 2\,3\,7\,5\,9\,0\,0\,0\,0\,0.$$

A simpler rule is to interpret E+09 as meaning "move the decimal point in 2.37590 nine places to the right."

[2] Suppose that the computer prints | 1.56056E–08 |

Here **E–08** means "multiply by 10 with **E**xponent **–8**."

Since $10^{-8} = \frac{1}{100{,}000{,}000} = 0.00000001$, 1.56056E–08 means

$$1.56056 \times 0.00000001 = 0.0\,0\,0\,0\,0\,0\,0\,1\,5\,6\,0\,5\,6.$$

A simpler rule is to interpret E–08 as meaning "move the decimal point in 1.56056 eight places to the left."

Now back to BLASTO. After studying your computer print-out, try these questions:

1. How fast will El Blasto be going after 1 second? Up or down?
2. How fast will El Blasto be going after 2 seconds? Up or down?
3. Which value of T in the FOR loop makes V closest to zero?
4. At about how many seconds after blast-out does El Blasto start to come down?
5. Can you estimate the total time for El Blasto to go from cannon to net? Does your estimate suggest that some of the values of V are physically impossible?

A good way to become familiar with the language of algebra is to investigate various scientific and business formulas for a large variety of values of the variables that they involve. The computer makes this easy, especially if we use the FOR . . . NEXT statement.

1C-2: CUBE If we are given that the side of a square or cube has a length of S centimeters, some formulas that we can investigate are:

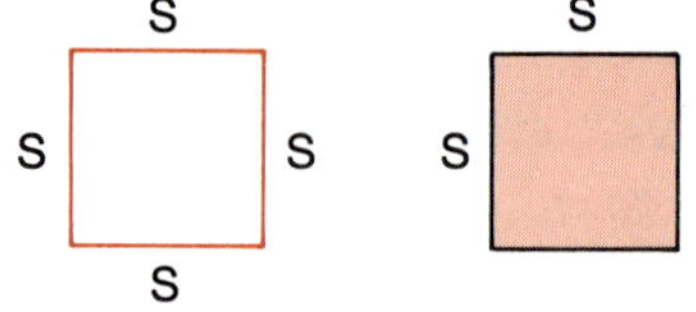

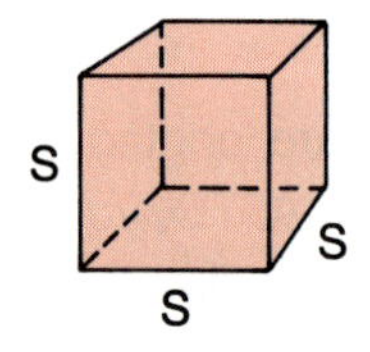

(a) Perimeter of the square (in centimeters):

$$P = S + S + S + S \text{ or } P = 4 * S$$

(b) Area of the square (in square centimeters): $A = S * S$

(c) Volume of a cube (in cubic centimeters): $V = S * S * S$

(d) Volume of a cube (in liters): $L = S * S * S * 0.001$ [1 liter = 1 cubic decimeter]

Here is a program to calculate the values for all the formulas given above for S = 1, 2, 3, 4, 5, 6, 7, 8, 9, and 10.

```
15  PRINT "SIDE (CM)","PER. (CM)","AREA (CM↑2)",
16  PRINT "VOL. (CM↑3)", "VOL. (L)"
20  FOR S=1 TO 10
30  LET P=4*S
40  LET A=S*S
50  LET V=S*S*S
60  LET L=.001*V
70  PRINT S,P,A,V,L
80  NEXT S
90  END
```

RUN this program and study the print-out. Notice that A increases *more* rapidly than S (we say that this is because area is proportional to the *square* of the side), while V increases *much more* rapidly than S (we say that volume is proportional to the *cube* of the side).

1C-3: SHRINK Rewrite program 1C-2 using only five BASIC statements.

1C-4: STRETCH Rewrite program 1C-3 to evaluate all five formulas for S = 0, 0.5, 1.0, 1.5, 2.0, . . . , N meters, where N is specified by the user. HINT: INPUT N before you print the headings, and change statement 20 in two ways.

1C-5: CHOICE Look in your library for a copy of the *Handbook of Chemistry and Physics* published by the Chemical Rubber Co., Cleveland, Ohio 44128. Write a program to calculate the values of some of the complicated looking formulas you find in this handbook, using various values of the variables.

Unit 1D. Decisions and Formulas

Suppose that you have a job that pays R dollars per hour. Then the formula for your weekly pay is

$$P = R * H$$

where H is the number of hours worked in a week. *But*—you may be working for a company that gives "time and a half" for overtime. This means that when you work more than 40 hours, the hourly rate becomes R * 1.5.

The formula for the weekly pay of a person who works more than 40 hours is:

$$P = \underbrace{R * 40}_{\text{Regular pay}} + \underbrace{R * 1.5 * (H - 40)}_{\text{Overtime pay}}$$

1D-1: PAYROLL Write a program in which you can input a worker's employee code (a number given each person to protect privacy), his hourly wage rate, and the number of hours he worked during the week. The program should then print out each employee code, the salary due, and the word "overtime" if the employee worked more than 40 hours. Typing in an employee code of "0" will make the program stop.

To analyze this program, a flow chart will help:

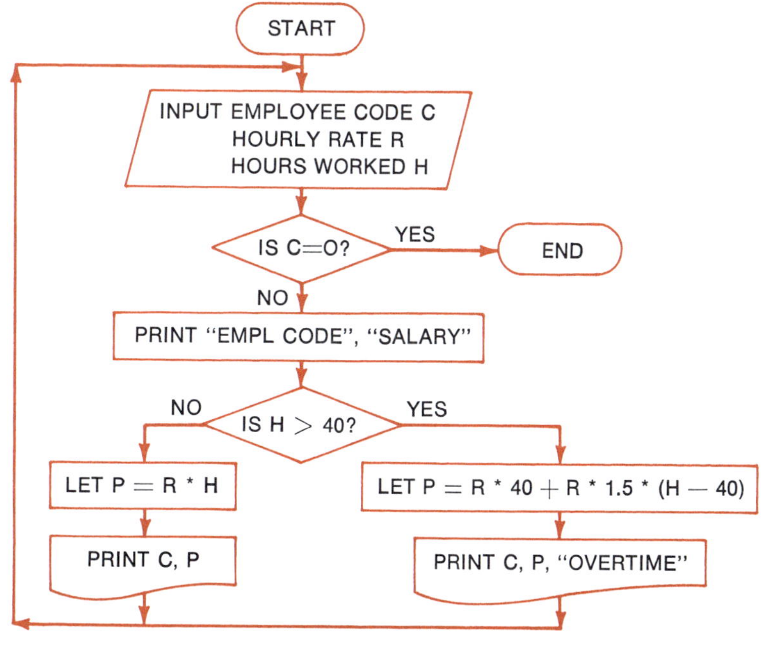

Here is a sample RUN:

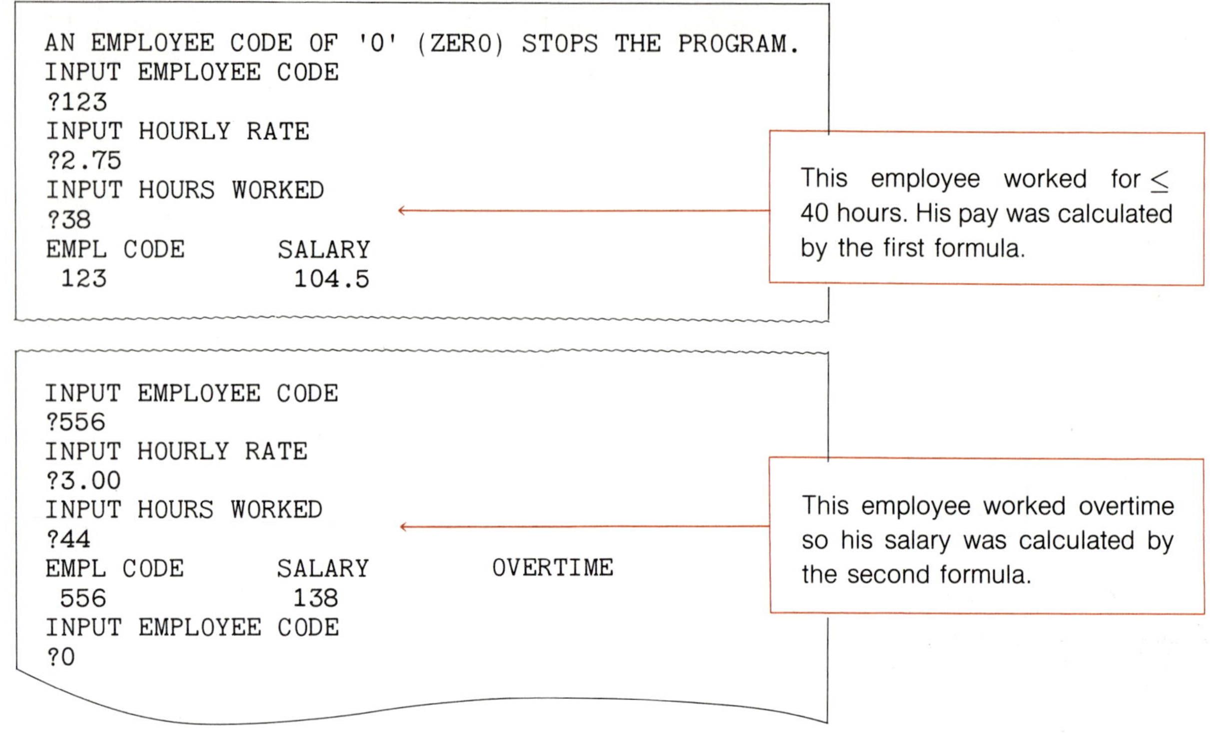

1D-2: TAX Program 1D-1 calculated the *gross* salary before deductions. Modify it to show a "tax deduction" based on the following table:

Weekly Salary	*Deduction*
$0 \leq P < 50.00$	5%
$50.00 \leq P < 100.00$	10%
$100.00 \leq P < 200.00$	15%
$200.00 \leq P$	20%

The output should include: EMPLOYEE CODE, GROSS PAY, DEDUCTION, NET PAY, OVERTIME.

1D-3: CHECK Modify Program 1D-2 so that it prints out "check reports" for each employee. Your output should be a sequence of reports to be cut between the dotted lines:

```
- - - - - - - - - - - - - - - - - - - - - - - - - - - - - - - - - - - - - - - - - - - -
- - - - - - - - - - - - - - - - - - - - - - - - - - - - - - - - - - - - - - - - - - - -

WEEK OF NOVEMBER 26, 1973
EMPLOYEE #032
REGULAR HOURS WORKED:  40    OVERTIME:   5
HOURLY RATE:   $ 2       GROSS PAY:   $ 95
"TAX" WITHHELD:   $ 9.5

NET PAY:   $ 85.5

- - - - - - - - - - - - - - - - - - - - - - - - - - - - - - - - - - - - - - - - - - - -
- - - - - - - - - - - - - - - - - - - - - - - - - - - - - - - - - - - - - - - - - - - -
```

And so on.

1D-4: IRS Write a program in which deductions are based on an actual deduction chart you have obtained from the Internal Revenue Service (look under U.S. Government in the telephone book for an address). If this kind of application interests you, you might talk to someone involved in the use of computers for payroll calculations and find out what other kinds of deductions must be made.

Unit 1E. Temperature Profiles

In Unit 1E, we will suggest some ways in which you can make the terminal "display" quantities on the number line in a pictorial manner.

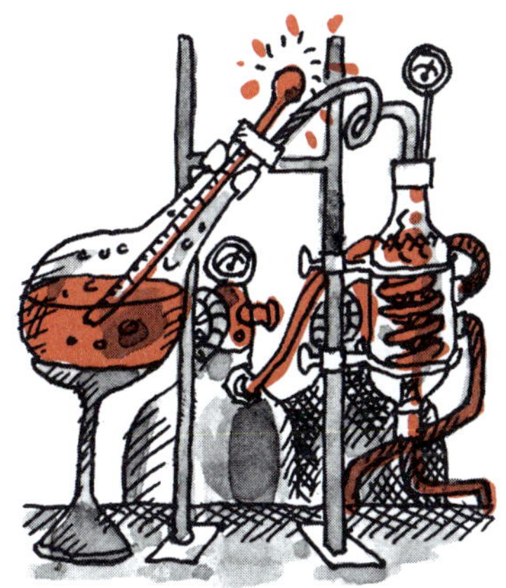

1E-1: COMPOUND-X3 A new chemical compound called X3 is very tempermental with respect to extreme temperatures. If the temperature gets too hot *or* too cold—horrible things happen to it. However, the *range* of temperatures over which compound X3 is safe can be adjusted at the lab for different customers (within limits). The program shown below produces an individual profile (picture) of the danger range of temperatures for each customer. Let's first look at a RUN of the program:

```
RUN


THIS PROGRAM DISPLAYS THE 'DANGER' AND
'SAFETY' TEMPERATURE ZONES OF A SPECIAL
ORDER OF COMPOUND X3.  DEGREES CENTIGRADE ARE USED.

WHAT IS THE LOWEST SAFE TEMPERATURE REQUESTED
(INPUT A NUMBER BETWEEN -50 AND +50)
?-30
WHAT IS THE HIGHEST TEMPERATURE REQUESTED
(INPUT A NUMBER BETWEEN -50 AND +50)
?20
THANK YOU--HERE IS THE DANGER-SAFETY PROFILE
FOR YOUR COMPOUND (D=DANGER; S=SAFETY)

-50  -40  -30  -20  -10    0  +10  +20  +30  +40  +50
--'----'----'----'----'----'----'----'----'----'----'--
DDDDDDDDDDDDSSSSSSSSSSSSSSSSSSSSSSSSSSDDDDDDDDDDDDDDDDD
```

A listing of our program is shown on the next page.

```
10  PRINT "THIS PROGRAM DISPLAYS THE 'DANGER' AND"
20  PRINT " 'SAFETY' TEMPERATURE ZONES OF A SPECIAL"
30  PRINT "ORDER OF COMPOUND X3. DEGREES CENTIGRADE ARE USED."
35  PRINT
40  PRINT "WHAT IS THE LOWEST SAFE TEMPERATURE REQUESTED"
60  PRINT "(INPUT A NUMBER BETWEEN -50 AND +50)"
70  INPUT L
75  IF L<-50 THEN 60
76  IF L>50 THEN 60
80  PRINT "WHAT IS THE HIGHEST TEMPERATURE REQUESTED"
90  PRINT "(INPUT A NUMBER BETWEEN -50 and +50)"
110  INPUT H
115  IF H<-50 THEN 90
116  IF H>50 THEN 90
120  PRINT "THANK YOU--HERE IS THE DANGER-SAFETY PROFILE"
130  PRINT "FOR YOUR COMPOUND (D=DANGER; S=SAFETY)"
140  PRINT
150  PRINT "-50   -40   -30   -20   -10     0   +10   +20   +30   +40   +50"
160  PRINT "--'----'----'----'----'----'----'----'----'----'----'--"
170  PRINT "D";
180  LET A=-50
190  PRINT "D";
200  LET  A=A+2
210  IF A <= L THEN 190
220  LET B=A
230  PRINT "S";
240  LET B=B+2
250  IF B <= H+2 THEN 230
260  LET C=B
270  PRINT "D";
275  LET C=C+2
280  IF C <= 50 THEN 270
290  PRINT "DDD"
295  PRINT
300  END
```

Check that the customer input was between -50 and $+50$. (lines 75–76)

Same check as above. (lines 115–116)

This is the tricky part! We'll let you figure it out. Just one help: we let $A = A + 2$ as a means of moving from left to right. The number 2 is used because our number scale has a mark for every 2 units. If enlarged, it would look like this: (lines 170–290)

```
-10   -8   -6   -4   -2    0   +2   +4   +6   +8   +10
  '    -    -    -    -    '    -    -    -    -     '
```

```
RUN

LOWEST? -20
HIGHEST? 40
THANK YOU
[+]  100  DANGER!
[+]   80  DANGER!
[+]   60  DANGER!
[+]   40  SAFE
[+]   20  SAFE
[+]   0   SAFE
[-]  -20  SAFE
[-]  -40  SAFE
[-]  -60  DANGER!
[-]  -80  DANGER!
```

1E-2: VERTICAL Write and RUN a program similar to program 1E-1 that prints the temperature scale *vertically*. You might also try making your program general enough to accept minimum-maximum limits different from $-50°$ to $+50°$. A sample run is pictured at the left. Try a more detailed scale if you wish.

1E-3: SNEAKY Can you trust a computer? Here's a program that might make you wonder. First look at the two RUNs on the next page.

```
RUN


THE COMPUTER HAS JUST SELECTED A LOWEST (MINIMUM)
SAFE TEMPERATURE (L) AND A HIGHEST (MAXIMUM) SAFE
TEMPERATURE (H).
NOW IT'S YOUR MOVE.  AT WHAT TEMPERATURE (SELECT
AN EVEN INTEGER BETWEEN -50 AND +50) WILL YOU STORE
COMPOUND X3?50
STAND BY....  HERE ARE THE RESULTS:
COMPUTER SELECTION:
                                   L= 5               H= 40
                                    \SSSSSSSSSSSSSSSS/
-50  -40  -30  -20  -10    0  +10  +20  +30  +40  +50
--'----'----'----'----'----'----'----'----'----'----'-
                                                     ↑
                                                     50
BOTTOM LINE WAS YOUR SELECTION...OUT OF SAFE RANGE

*************          SORRY

BBBB    000    000   M     M
B   B  0   0  0   0  MM   MM
B   B  0   0  0   0  M M M M
BBBB   0   0  0   0  M  M  M
B   B  0   0  0   0  M     M
B   B  0   0  0   0  M     M
BBBB    000    000   M     M
```

```
RUN


THE COMPUTER HAS JUST SELECTED A LOWEST (MINIMUM)
SAFE TEMPERATURE (L) AND A HIGHEST (MAXIMUM) SAFE
TEMPERATURE (H).
NOW IT'S YOUR MOVE.  AT WHAT TEMPERATURE (SELECT
AN EVEN INTEGER BETWEEN -50 AND +50) WILL YOU STORE
COMPOUND X3?-10
STAND BY....  HERE ARE THE RESULTS:
COMPUTER SELECTION:
                                L= 0   H= 15
                                 \SSSSSS/
-50  -40  -30  -20  -10    0  +10  +20  +30  +40  +50
--'----'----'----'----'----'----'----'----'----'----'-
                        ↑
                       -10
BOTTOM LINE WAS YOUR SELECTION...OUT OF SAFE RANGE

*************          SORRY

BBBB    000    000   M     M
B   B  0   0  0   0  MM   MM
B   B  0   0  0   0  M M M M
BBBB   0   0  0   0  M  M  M
B   B  0   0  0   0  M     M
B   B  0   0  0   0  M     M
BBBB    000    000   M     M
```

Here's a listing of the program that produced those RUNs on page 37.

```
100  PRINT "THE COMPUTER HAS JUST SELECTED A LOWEST (MINIMUM)"
110  PRINT "SAFE TEMPERATURE (L) AND A HIGHEST (MAXIMUM) SAFE"
120  PRINT "TEMPERATURE (H)."
130  PRINT "NOW IT'S YOUR MOVE.  AT WHAT TEMPERATURE (SELECT"
140  PRINT "AN EVEN INTEGER BETWEEN -50 AND +50) WILL YOU STORE"
150  PRINT "COMPOUND X3";
160  INPUT X
170  PRINT "STAND BY....  HERE ARE THE RESULTS:"
180  PRINT "COMPUTER SELECTION:"
190  IF X<0 THEN 230
200  LET H=X-10
210  LET L=H-(X/2+10)
220  GOTO 250
230  LET L=X+10
240  LET H=L+(-X/2+10)
250  LET J1=27+(L/10*5)
260  LET J2=27+(H/10*5)
270  PRINT TAB(J1-1);"L=";L;TAB(J2-1);"H=";H
280  PRINT TAB(J1);"\";
290  FOR I=J1+1 TO J2-1
300  PRINT "S";
310  NEXT I
320  PRINT "/"
330  PRINT "-50  -40  -30  -20  -10    0  +10  +20  +30  +40  +50"
340  PRINT "--'----'----'----'----'----'----'----'----'----'----'-"
350  PRINT TAB(27+(X/10*5));"↑"
360  PRINT TAB(26+(X/10*5));X
370  PRINT "BOTTOM LINE WAS YOUR SELECTION...";
380  PRINT "OUT OF SAFE RANGE"
390  PRINT
400  PRINT "*************";TAB(20);"SORRY"
410  PRINT
420  PRINT "BBBB    OOO    OOO  M     M"
430  PRINT "B   B  O   O  O   O MM   MM"
440  PRINT "B   B  O   O  O   O M M M M"
450  PRINT "BBBB   O   O  O   O M  M  M"
460  PRINT "B   B  O   O  O   O M     M"
470  PRINT "B   B  O   O  O   O M     M"
480  PRINT "BBBB    OOO    OOO  M     M"
490  END
```

Even if you ran this program hundreds of times, you could never select a safe temperature. The reason is that the computer cheats! It really doesn't select L and H until *after* you input the temperature at which you wish to store compound X3. Then in a flash (computers are very fast), the program quickly selects a 'safe' range SSSSSS that is to the *left* if you input a positive temperature or to the *right* if you input a negative temperature. To look convincing, the program varies the range of safe temperatures from \SSSS/ to \SSSSSSSSSSSSSSS/.

The rules that were used in Program 1E-3 to make sure that the computer always wins the game are:

1. Place the user's input in location X.

2. If X > 0, select a 'safe' range to the *left* of X, with the high end (H) of the safe range at 10 degrees below X. The lower end (L) is set as H − (X/2 + 10). For example, suppose that the student selects a temperature X = 40; then the computer sets:

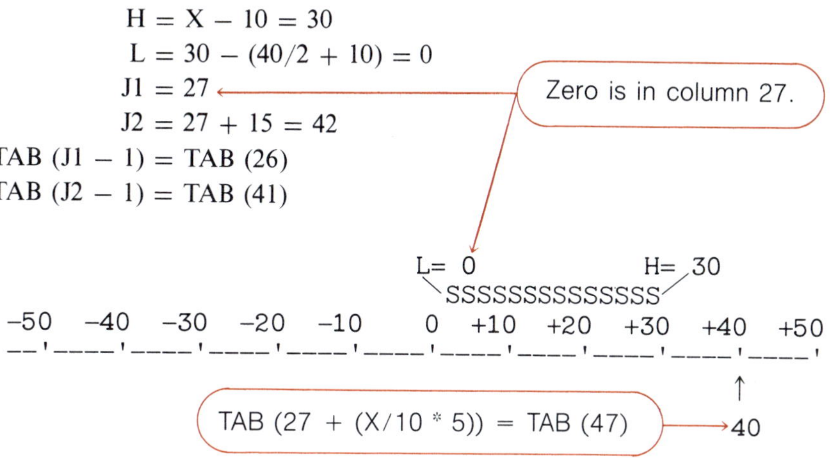

3. If X < 0, use a similar rule to place the safe range on the *right*.

1E-4: HONEST (Don't try this unless you are willing to read ahead to section 4 to see what a random generator is.)

Rewrite the above program to make it "honest," using the random generator RND to select the values of L and H. If the temperature selected by the user falls between these numbers, print a congratulatory message; otherwise KERBOOM! HINT: Try this test program:

```
10  FOR K=1 TO 50
20  PRINT INT(101*RND(1)-50),
30  NEXT K
40  END
```

In general, the formula N=INT((S−R+1)*RND(1)+R) produces randomly selected integers from R to S inclusive. If we let R = −50 and S = +50, then S − R = 100. This is how we got the formula in line 20 above.

SECTION 2 OPERATIONS WITH REAL NUMBERS

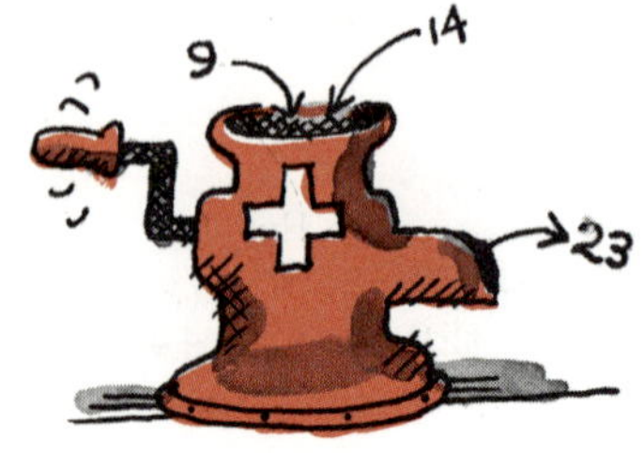

Checklist of Computing Skills

Previously explained:	PRINT, END, LET, INPUT, IF . . . THEN, STOP GOTO, FOR, NEXT, ABS, INT, SQR
Explained in this section:	GOTO K OF . . . (or ON K GOTO . . .)

Checklist of Algebraic Skills

1. $(-7) - (-8) =$ __?__
2. $(-3) * (-8) =$ __?__
3. $(+6) - (2) * (-3) =$ __?__
4. $(-8) + (2) + (-3) - (-7) + (-10) - (-1) =$ __?__
5. The average value of the six numbers in Ex. 4 = __?__

When you have finished, check your answers with those at the bottom of the page.

NEW: GOTO K OF . . . (or ON K GOTO . . .)

GOTO K OF and ON K GOTO are different forms (used by different computers) of the "multiple branch" statement. For example,

GOTO K OF 17,45,80

(or ON K GOTO 17,45,80) means that the program should GOTO line number 17 IF K = 1, to line 45 IF K = 2, and to line 80 IF K = 3. K is a variable whose value is determined by the program.

If K happens to be a decimal, then some computers use the "integer part" of the number stored in K. However, the Time Share Corporation system uses the *rounded* value of the variable. In either case, if the value of K is *none* of the numbers 1, 2, or 3, then the program simply proceeds to the next step (default condition).

Study this example; also try it on your computer, which *may* act differently.

```
10  INPUT K
20  GOTO K OF 50,70,90
30  PRINT "DEFAULT BRANCH"
40  GOTO 10
50  PRINT "BRANCH FOR K=1"
60  GOTO 10
70  PRINT "BRANCH FOR K=2"
80  GOTO 10
90  PRINT "BRANCH FOR K=3"
100  GOTO 10
110  END
RUN
```

Your computer may use:
20 ON K GOTO 50,70,90

Answers: (1) 1 (2) 24 (3) 12 (4) −11 (5) −1.8333...

```
?-2
DEFAULT BRANCH
?0
DEFAULT BRANCH
?.45
DEFAULT BRANCH
?.55
BRANCH FOR K=1
?2
BRANCH FOR K=2
?3.1416
BRANCH FOR K=3
?3.8
DEFAULT BRANCH
?4
DEFAULT BRANCH
?
END
```

There can be more (or less) than 3 branches. Here's an example with 4 branches for you to study and RUN:

```
10  PRINT "QUIZ:  TYPE THE NUMBER OF THE CORRECT ANSWER."
20  PRINT "YOU CAN EXAMINE AN OBELISK IF YOU VISIT:"
30  PRINT "(1) WASHINGTON, D. C.     (2) YOUR DOCTOR"
40  PRINT "(3) THE LIBRARY           (4) THE TEMPLE OF KARNAK"
50  PRINT "NOW WHAT IS YOUR ANSWER";
60  INPUT A
70  GOTO A OF 100,200,300,400
80  PRINT "READ THE INSTRUCTIONS AGAIN."
90  GOTO 50
100  PRINT "RIGHT--IT'S THE WASHINGTON MONUMENT.  (4 IS ALSO CORRECT.)"
110  STOP
200  PRINT "NOT UNLESS HE COLLECTS STRANGE PAPER WEIGHTS."
210  STOP
300  PRINT "THEY'LL PROBABLY ONLY HAVE PICTURES."
310  STOP
400  PRINT "EXCELLENT!  (1 IS ALSO CORRECT.)"
410  END
```

$-3-2=-5$

$+5-4=+1$

$-4+3=-1$

$-3(-2)=+6$

$+2(-4)=-8$

Unit 2A. Operations with Positive and Negative Numbers

Understanding the rules for operations on numbers illustrated in our diagram is one kind of skill that comes out of a study of algebra; being able to *quickly* give answers to such problems (almost without thinking) is a different, but equally valuable skill. This unit shows you how to write programs that coach you at being expert at the "almost without thinking" skill.

2A-1: ARITH This program gives practice in adding, subtracting, and multiplying integers. The best way to understand how the program works is to compare the flow chart below with the actual program listing on the next page.

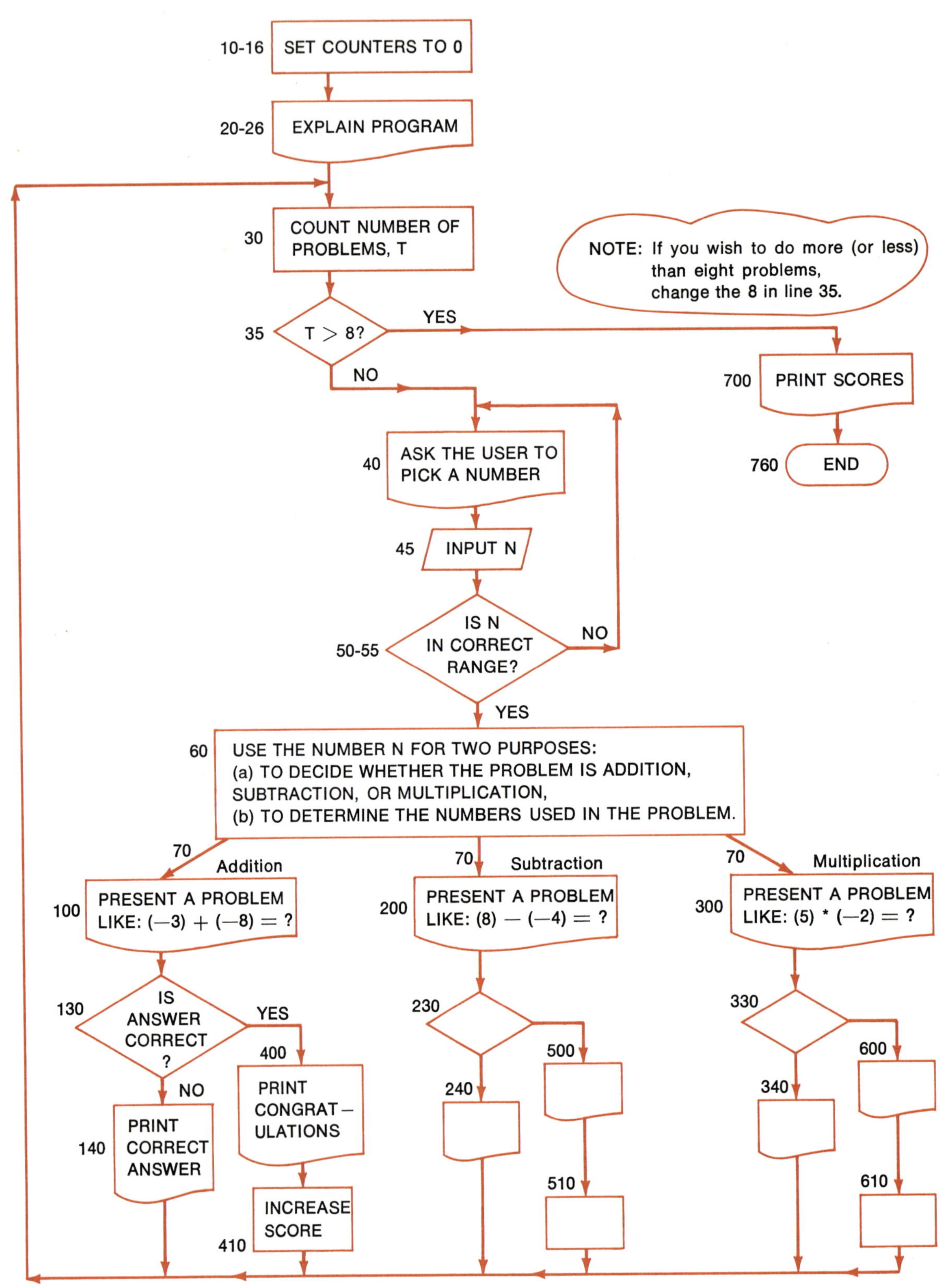

Here is a listing of a program based on the flow chart:

```
10  LET T=0
12  LET A=0
14  LET S=0
16  LET M=0
20  PRINT "THIS PROGRAM WILL PRESENT YOU WITH 8 PROBLEMS."
22  PRINT "THE PROBLEMS WILL BE IN ADDITION, SUBTRACTION,"
24  PRINT "    AND MULTIPLICATION."
26  PRINT "THE PROBLEM IS DETERMINED BY AN INTEGER YOU SELECT."
30  LET T=T+1
35  IF T>8 THEN 690
38  PRINT
40  PRINT "TYPE AN INTEGER BETWEEN -10 AND 10."
45  INPUT N
50  IF N <= -10 THEN 40
55  IF N >= 10 THEN 40
60  LET K=ABS(INT(N/3))
70  GOTO K OF 100,200,300
80  GOTO 300
100  PRINT "ADDITION PROBLEM: ";
110  PRINT "(";N;") + (";N-5;") = ";
120  INPUT X
130  IF X=2*N-5 THEN 400
140  PRINT "NO, THE ANSWER IS ";2*N-5;"."
150  GOTO 30
200  PRINT "SUBTRACTION PROBLEM: ";
210  PRINT "(";N;") - (";2*N-20;") = ";
220  INPUT X
230  IF X=20-N THEN 500
240  PRINT "NO, THE ANSWER IS ";20-N;"."
250  GOTO 30
300  PRINT "MULTIPLICATION PROBLEM: ";
310  PRINT "(";N;") * (";N-7;") = ";
320  INPUT X
330  IF X=N*(N-7) THEN 600
340  PRINT "NO, THE ANSWER IS ";N*(N-7);"."
350  GOTO 30
400  PRINT "CORRECT--THAT'S A POINT FOR YOU."
410  LET A=A+1
420  GOTO 30
500  PRINT "EXCELLENT--CHALK UP A CORRECT SUBTRACTION."
510  LET S=S+1
520  GOTO 30
600  PRINT "WOW!  YOU DID A MULTIPLICATION CORRECTLY."
610  LET M=M+1
620  GOTO 30
690  PRINT
700  PRINT "O.K.  HERE ARE YOUR FINAL SCORES:"
710  PRINT "YOU HAD: ";A;" RIGHT IN ADDITION."
720  PRINT "          ";S;" RIGHT IN SUBTRACTION."
730  PRINT "          ";M;" RIGHT IN MULTIPLICATION."
740  PRINT "YOU HAD A TOTAL OF";A+S+M;" RIGHT---GRADE =";
745  PRINT (A+S+M)/8*100;"%"
750  PRINT "THAT'S ALL FOR NOW.  BYE."
760  END
```

The trick we use to get K in line 60 is to take the value of N the user inputs, divide it by 3, take the integer part, and then take the absolute value. For example, if $N = -7$, $N/3 = -2.33$, $INT(-2.33) = -3$, $ABS(-3) = 3$. Thus $K = 3$, which causes a GOTO 300.

Notice that line 70 causes a 3-way branch. If $K = 1$, the branch is to line 100; if $K = 2$, to line 200; and if $K = 3$, to line 300. Any other value of K causes the statement after 70 (line 80) to be executed.

Here is a sample run of program 2A-1:

```
RUN


THIS PROGRAM WILL PRESENT YOU WITH 8 PROBLEMS.
THE PROBLEMS WILL BE IN ADDITION, SUBTRACTION,
   AND MULTIPLICATION.
THE PROBLEM IS DETERMINED BY AN INTEGER YOU SELECT.

TYPE AN INTEGER BETWEEN -10 AND 10.
?4
ADDITION PROBLEM: ( 4) + (-1) = ?3
CORRECT--THAT'S A POINT FOR YOU.

TYPE AN INTEGER BETWEEN -10 AND 10.
?8
SUBTRACTION PROBLEM: ( 8) - (-4) = ?12
EXCELLENT--CHALK UP A CORRECT SUBTRACTION.

TYPE AN INTEGER BETWEEN -10 AND 10.
?0
MULTIPLICATION PROBLEM: ( 0) * (-7) = ?0
WOW!  YOU DID A MULTIPLICATION CORRECTLY.

TYPE AN INTEGER BETWEEN -10 AND 10.
?10
TYPE AN INTEGER BETWEEN -10 AND 10.
?9
MULTIPLICATION PROBLEM: ( 9) * ( 2) = ?18
WOW!  YOU DID A MULTIPLICATION CORRECTLY.

TYPE AN INTEGER BETWEEN -10 and 10.
?-2
ADDITION PROBLEM: (-2) + (-7) = ?-9
CORRECT--THAT'S A POINT FOR YOU.

TYPE AN INTEGER BETWEEN -10 AND 10.
?-8
MULTIPLICATION PROBLEM: (-8) * (-15) = ?120
WOW!  YOU DID A MULTIPLICATION CORRECTLY.

TYPE AN INTEGER BETWEEN -10 AND 10.
?-5
SUBTRACTION PROBLEM: (-5) - (-30) = ?25
EXCELLENT--CHALK UP A CORRECT SUBTRACTION.

TYPE AN INTEGER BETWEEN -10 AND 10.
?7
SUBTRACTION PROBLEM: ( 7) - (-6) = ?13
EXCELLENT--CHALK UP A CORRECT SUBTRACTION.

O.K.  HERE ARE YOUR FINAL SCORES:
YOU HAD:  2 RIGHT IN ADDITION.
          3 RIGHT IN SUBTRACTION.
          3 RIGHT IN MULTIPLICATION.

YOU HAD A TOTAL OF 8 RIGHT---GRADE = 100%
THAT'S ALL FOR NOW. BYE.
```

Here are two questions to see how good a program analyst you are:

(a) For what values of K will you get a multiplication problem?
(b) For what values of N will you get a multiplication problem?

NOTE: Since the preceding program is rather long, some classes may decide to have one student enter the program, and then leave it in the computer's memory long enough for everyone to run it. Classes should also save long programs of this kind on paper tape (or on magnetic tape or disc) for future use. The book *A Guided Tour of Computer Programming in BASIC* explains the use of paper tape on pages 78 to 82. The reference manual for your computer should also be consulted.

2A-2: HARDARITH The "data" for the problems in program 2A-1 were all generated from the number N entered by the person running the program. For example, multiplication problems were given in line 310 as:

$$N * (N - 7)$$

The choice of -7 was arbitrary. Notice that it caused different signs to appear in the problems. For example,

$N = 9$ caused the problem to be $(9) * (2)$
$N = 0$ caused the problem to be $(0) * (-7)$
$N = -8$ caused the problem to be $(-8) * (-15)$

For program 2A-2, change lines 110, 210, and 310 so that they produce "harder" problems. WARNING: You must also change the lines that check for correct answers (lines 130, 140, 230, 240, 330, 340).

2A-3: DIVISION Modify program 2A-1 so that it also gives problems in division.

If you still find these programs too easy, try using a decimal number instead of an integer as the "selection number." (You may then need paper and pencil to do the calculations.) Change lines 26 and 40 in the program.

Unit 2B. Approximate Arithmetic

Can you scan the contents of a basket full of groceries at the supermarket and estimate the total bill *without* using paper and pencil?

The practice program in this unit is meant to give you experience in developing such "amazing" powers of mental calculation. The ability to *estimate* sums and products is one of the most useful skills you can develop. You will find that you can invent your own methods after a while and become something of a "soothsayer" in matters mathematical.

Here's an example of one "trick" you can use for estimating sums of grocery item prices. First estimate the number of half-dollars; then add:

Price	*Approximate No. of Half-dollars*	
1.69	3^+	You might think:
.99	2	$3^+ + 2 = 5^+$
.49	1	$5^+ + 1 + 1^- = 7$
.17	1^-	$7 + 1^+ + 2 = 10^+$
.72	1^+	
1.05	2	
$5.11	10^+	

What you do is convert each price into a number of "half dollars." You can use a + (or −) to indicate that the price is actually above (or below) the number of half dollars you chose. As you go down the list, use the rule that each $^-$ cancels out a $^+$. At the end, each + (or −) sign left over represents roughly + or − a quarter. So in the above example, our approximate answer is 10 half dollars plus one quarter, or $5.25 (which misses by 14 cents which is an error of only about 2.7%).

Try other methods. For a rougher estimate, try rounding each price to the the nearest dollar. For a closer estimate, try rounding each price to the nearest number of dimes. In our example, the rounded number of dimes would be

$$17 + 10 + 5 + 2 + 7 + 10 = 51 \text{ or approximately } \$5.10.$$

2B-1: WHIZADD Here is a run of a computer program that allows you to practice estimating sums. It gives you both the correct answer and the percentage by which your estimate missed the exact sum.

```
PRACTICE IN DEVELOPING 'SUPER-SPEED' ADDITION SKILL.
ANSWERS WITHIN 10% ARE MARKED CORRECT.
HOW MANY PROBLEMS DO YOU WANT TO TRY?4
TYPE IN THE TIME (HOUR,MINUTES):?12,58
PROBLEM  1
 3.11  +  8.71  +  3.91  +  4.91  +  0.41  +  0.41  =
 ?21.00
VERY GOOD
CORRECT ANSWER WAS  21.46  (YOU WERE WITHIN  2.14352 %)

PROBLEM  2
 6.63  +  5.19  +  0.39  +  8.43  +  7.45  +  3.11  =
 ?28.00
NOT TOO GOOD
CORRECT ANSWER WAS  31.2  (YOU WERE WITHIN  10.2564 %)
```

```
PROBLEM  3
 9.52  +  2.3  +  2.5  +  11.32  +  13.23  +  6  =
 ?45.00
VERY GOOD
CORRECT ANSWER WAS  44.87  (YOU WERE WITHIN  0.289725 %)

PROBLEM  4
 12.83  +  1.01  +  5.81  +  14.63  +  19.85  +  9.31  =
 ?64.00
VERY GOOD
CORRECT ANSWER WAS  63.44  (YOU WERE WITHIN  0.882723 %)

TYPE IN THE TIME (HOUR,MINUTES):?1,02
YOUR SCORE:
NUMBER OF ANSWERS WITH BETTER THAN 10% ACCURACY:  3
NUMBER OF ANSWERS WITH WORSE THAN 10% ACCURACY:  1
YOU'VE BEEN RUNNING THIS PROGRAM FOR  0  HOURS
AND  4  MINUTES: AVERAGE TIME PER PROBLEM WAS  60  SEC.
```

To make program 2B-1 interesting, we've used devious tricks to get "surprise" values for the 6 numbers called N1, N2, N3, H4, N5, and N6. Let's look at the way we get N2 to illustrate our devious thinking. Line 130 says (see next page):

```
130  LET N2=ABS(2.91+M1/10-K)
```

The ABS is just to make sure we only get positive numbers.

- The 2.91 is arbitrary—you could put any number you wanted here. We chose one to look like a price in dollars and cents.
- M1 is the number of minutes the student types in when asked the time. This will make the problems different when run at different times. Dividing by 10 keeps the value $< 60/10 = 6$.
- K is our surprise element. Here's how we get it. In line 420 we take the student's answer (A), and the correct answer (C), and calculate the *percent* by which they differ (E). For example, suppose that A = 4.55 and C = 4.70. Then

$$E = ABS(4.55 - 4.70)/4.70 * 100 = 0.15/4.70 * 100 = 3.19148$$

Now for our surprise number, K, we have decided to use

the second, third, and fourth decimal digits of E.

$$E = 3.19148$$

That's tricky to do—here's the technique:

First, we move the decimal 1 place to the right:	10 * E = 31.9148
Second, we take the integral part of this:	INT(10 * E) = 31
Third, we subtract:	0.9148
Fourth, we multiply by 1000:	914.8
Fifth, we take the integer part of this:	914
Sixth, we divide by 100:	9.14

And behold—we have a surprise number in the form of "dollars and cents."

If you found the above complicated, you're right! (There is a much easier way of getting surprise numbers in BASIC which will be explained in section 4.) Even if you don't fully understand this program, go ahead and type it in for use as a drill in mental arithmetic.

Here is a listing of program 2B-1:

```
10  LET K=0
20  LET R=0
30  LET W=0
40  PRINT "PRACTICE IN DEVELOPING 'SUPER-SPEED' ADDITION SKILL."
60  PRINT "ANSWERS WITHIN 10% ARE MARKED CORRECT."
70  PRINT "HOW MANY PROBLEMS DO YOU WANT TO TRY";
80  INPUT T
90  PRINT "TYPE IN THE TIME (HOUR,MINUTES):";
100  INPUT H1,M1
110  FOR I=1 TO T
120  LET N1=1.91+H1/10+K
130  LET N2=ABS(2.91+M1/10-K)
140  LET N3=ABS(3.91-K)
150  LET N4=4.91+K
160  LET N5=.41+2*K
170  LET N6=ABS(.41-K)
270  PRINT "PROBLEM ";I
280  PRINT N1;"  +  ";N2;"  +  ";N3;"  +  ";N4;"  +  ";N5;"  +  ";N6;"  ="
290  LET C=N1+N2+N3+N4+N5+N6
300  INPUT A
310  IF ABS(A-C)/C>.10 THEN 460
380  LET R=R+1
390  PRINT "VERY GOOD"
400  PRINT "CORRECT ANSWER WAS ";C;" (YOU WERE WITHIN ";
410  PRINT ABS(A-C)/C*100;"%.)"
420  LET E=ABS(A-C)/C*100
422  LET K=INT(1000*(10*E-INT(10*E)))/100
450  GOTO 500
460  LET W=W+1
470  PRINT "NOT TOO GOOD"
480  GOTO 400
500  PRINT
510  NEXT I
520  PRINT "TYPE IN THE TIME (HOUR,MINUTES):";
530  INPUT H2,M2
540  PRINT "YOUR SCORE:"
550  PRINT "NUMBER OF ANSWERS WITH BETTER THAN 10% ACCURACY: ";R
560  PRINT "NUMBER OF ANSWERS WITH WORSE THAN 10% ACCURACY: ";W
580  IF M2>M1 THEN 610
590  LET M=60-M1+M2
595  LET H2=H2-1
600  GOTO 620
610  LET M=M2-M1
620  IF H2>H1 THEN 650
630  LET H=12-H1+H2
640  GOTO 660
650  LET H=H2-H1
660  PRINT "YOU'VE BEEN RUNNING THIS PROGRAM FOR ";H;" HOURS"
670  PRINT "AND ";M;" MINUTES: AVERAGE TIME PER PROBLEM WAS ";
675  PRINT (H*3600+M*60)/T;" SEC."
680  END
```

Unit 2C. The MINICALC Simulator

Computers are called "general purpose" machines because they can be made to act like (or *simulate*) other devices. In the A and B units in this book, for example, you have been writing programs which simulate a coach or tutor. Computers have been used to simulate everything from elevator systems to the stock market.

In this unit we are going to write a program that makes the computer simulate (act like) a very simple desk calculator that can only *add* or *multiply*. Computers can do much more than this, of course, but there is more to the desk calculator simulator than meets the eye. In fact you'll have an important use for it in Unit 3B.

2C-1: CALC-1 Let's first write a program in BASIC which simulates a desk calculator that can *add* any number of terms, or *multiply* any number of factors. We'll assume that the numbers to be added or multiplied are less than 9,000,000,000. In computer language 9,000,000,000 is written in shorthand as 9E9 which means $9 * 10^9$ which is $9 \times 1{,}000{,}000{,}000 = 9{,}000{,}000{,}000$. This limit on number size allows us to interpret 9E9 as a special signal rather than a number to be added.

Let's see how this works by first writing a program that can only add numbers.

```
5  LET S=0
10  PRINT "INPUT THE NUMBERS TO BE ADDED."
20  PRINT "TYPE 9E9 WHEN YOU WISH TO STOP."
30  INPUT S1
40  IF S1 >= 9E9 THEN 70
50  LET S=S+S1
60  GOTO 30
70  PRINT "SUM IS ";S
80  END
```

To see how this works, we can compare a RUN with what happens to the variable S:

```
RUN


INPUT THE NUMBERS TO BE ADDED.
TYPE 9E9 WHEN YOU WISH TO STOP.
?25
?33
?-6
?10
?9E9
SUM IS 62
```

$S = 0$

$S = 0 + 25 = 25$
$S = 25 + 33 = 58$
$S = 58 + (-6) = 52$
$S = 52 + 10 = 62$
This is >= 9E9 so the program branches to 70.

You can see that the trick to adding a list of numbers is to repeatedly execute the statement LET S = S + S1 where S1 is the latest number to be added, and S is a variable that accumulates the sum. S must start as zero. (Why?)

For multiplication, the trick is to repeatedly execute the statement LET M = M * M1 where M1 is the latest number to be multiplied, and M is a variable that accumulates the product. M must start as 1. (Why?)

Here's what a short multiplication program looks like. Run this program with factors 33, 3, and 24.

```
5  LET M=1
10  PRINT "INPUT THE FACTORS TO BE MULTIPLIED."
20  INPUT M1
30  IF M1>=9E9 THEN 60
40  LET M=M*M1
50  GOTO 20
60  PRINT "PRODUCT IS";M
70  END
```

Now let's write a program where the user can give a "code" number to specify the operation desired. We will allow either addition (code 1) or multiplication (code 2) to be done over and over until you say "stop" (code 0). Line 177 protects against an illegal code like 3, 4, etc.

```
100  PRINT "MINICALC PROGRAM."
110  PRINT "THE AVAILABLE OPERATION CODES ARE:"
120  PRINT "    STOP                 0"
125  PRINT "    ADD                  1"
130  PRINT "    MULTIPLY             2"
150  PRINT "TYPE 0,1, OR 2 FOR AN OPERATION CODE."
160  INPUT C
170  IF C=0 THEN 700
175  IF C=2 THEN 300
177  IF C<>1 THEN 150
200  PRINT "INPUT THE NUMBERS TO BE ADDED, ONE AT A TIME."
210  PRINT "WHEN FINISHED TYPE 9E9."
215  LET S=0
220  INPUT S1
230  IF S1>=9E9 THEN 260
240  LET S=S+S1
250  GOTO 220
260  PRINT "THE SUM IS ";S
270  GOTO 150
300  PRINT "INPUT THE FACTORS TO BE MULTIPLIED, ONE AT A TIME."
310  PRINT "WHEN FINISHED TYPE 9E9."
315  LET M=1
320  INPUT M1
330  IF M1>=9E9 THEN 360
340  LET M=M*M1
350  GOTO 320
360  PRINT "THE PRODUCT IS ";M
370  GOTO 150
700  PRINT "END OF MINICALC PROGRAM."
710  END
```

ADDITION (lines 200–270)

MULTIPLICATION (lines 300–370)

Let's try the preceding program on the following problem:

What will it cost you to buy a lamp at \$33.49, a radio at \$17.78, and a chair at \$54.23 if the dealer allows a 15% discount, but there is a 6% sales tax in your state?

HINT 1: Instead of taking 15% discount and subtracting, it's faster to take 85% of the list price, since:

$$P - .15 * P = .85 * P$$

Proof: $P - .15 * P = 1 * P - .15 * P = (1.00 - 0.15) * P = 0.85 * P$

HINT 2: It is smarter to add first and *then* multiply, since $.85 * A + .85 * B + .85 * C = .85 * (A + B + C)$. (Distributive Law)

HINT 3: The fastest way to add in the sales tax is to multiply the discount price by 1.06, since $D + .06 * D = 1.06 * D$.

```
RUN


MINICALC PROGRAM.
THE AVAILABLE OPERATION CODES ARE:
   STOP                 0
   ADD                  1
   MULTIPLY             2
TYPE 0,1, OR 2 FOR AN OPERATION CODE.
?1
INPUT THE NUMBERS TO BE ADDED, ONE AT A TIME.
WHEN FINISHED TYPE 9E9.
?33.49
?17.78
?54.23
?9E9
THE SUM IS 105.5
TYPE 0,1, OR 2 FOR AN OPERATION CODE.
?2
INPUT THE FACTORS TO BE MULTIPLIED, ONE AT A TIME.
WHEN FINISHED TYPE 9E9.
?105.5
?.85
?9E9
THE PRODUCT IS 89.675
TYPE 0,1, OR 2 FOR AN OPERATION CODE.
?2
INPUT THE FACTORS TO BE MULTIPLIED, ONE AT A TIME.
WHEN FINISHED TYPE 9E9.
?89.68
?1.06
?9E9
THE PRODUCT IS 95.0608
TYPE 0,1, OR 2 FOR AN OPERATION CODE.
?0
END OF MINICALC PROGRAM.
```

TOTAL LIST PRICE (arrow to "THE SUM IS 105.5")

DISCOUNT PRICE (arrow to "THE PRODUCT IS 89.675")

DISCOUNT PRICE WITH TAX ADDED (arrow to "THE PRODUCT IS 95.0608")

2C-2: DECIDE Rewrite the above program so that the user inputs the number of terms to be added (or the number of factors to be multiplied). In this way the code 9E9 will not have to be used. (HINT: Use FOR . . . NEXT loops.)

Unit 2D. The MIDICALC Simulator

The picture below suggests that calculators can be made more useful by adding other operations. For most applications, *division* is the most useful new feature that might be added to the "mini" machine to make a "midi" machine. In this unit we'll add the division operation to our simulated calculator.

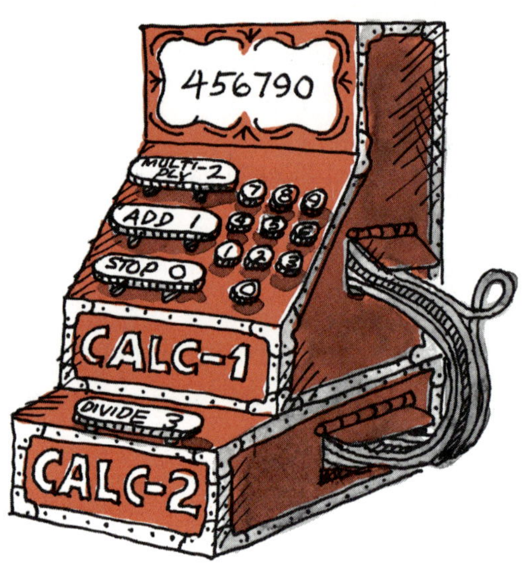

2D-1: CALC-2 Program 2C-1 allowed the user to input a long sequence of terms to be added, or a long sequence of factors to be multiplied. It didn't matter in what *order* this was done, because both the addition and multiplication of real numbers are *commutative* operations. In other words:

$$4 + 8 = 8 + 4 \qquad \text{and} \qquad (6) * (-2) = (-2) * (6)$$

It was also O.K. to put in more than two numbers at a time, since both of these operations are *associative,* that is:

$$(4 + 8) + 2 = 4 + (8 + 2) \qquad \text{and} \qquad ((6) * (-2)) * 3 = 6((-2) * (3))$$

But . . .

Division of real numbers is neither commutative nor associative.

Noncommutative Example:	$15/3 = 5$ but $3/15 = 0.2$
Nonassociative Example:	$(8/4)/2 = (2)/2 = 1$
but	$8/(4/2) = 8/(2) = 4$

This means that our CALC-2 division program can only ask the user for *two* numbers at a time, and it will also have to ask which one is the *dividend,* and which one is the *divisor*. Here's a sample RUN:

```
RUN


MIDICALC PROGRAM.
THE AVAILABLE OPERATION CODES ARE:
   STOP                   0
   ADD                    1
   MULTIPLY               2
   DIVIDE                 3
TYPE 0,1,2,OR 3 FOR AN OPERATION CODE.
?3
INPUT THE DIVIDEND?99
INPUT THE DIVISOR?66
THE QUOTIENT IS  1.5
TYPE 0,1,2,OR 3 FOR AN OPERATION CODE.
?3
INPUT THE DIVIDEND?89
INPUT THE DIVISOR?44
THE QUOTIENT IS  2.02273
TYPE 0,1,2,OR 3 FOR AN OPERATION CODE.
?2
INPUT THE FACTORS TO BE MULTIPLIED, ONE AT A TIME.
WHEN FINISHED TYPE 9E9.
?56
?89
?9E9
THE PRODUCT IS  4984
TYPE 0,1,2,OR 3 FOR AN OPERATION CODE.
?0
END OF MIDICALC PROGRAM.
```

Here is a listing of the program 2D-1:

```
100  PRINT "MIDICALC PROGRAM."
110  PRINT "THE AVAILABLE OPERATION CODES ARE:"
120  PRINT "   STOP                   0"
125  PRINT "   ADD                    1"
130  PRINT "   MULTIPLY               2"
135  PRINT "   DIVIDE                 3"
150  PRINT "TYPE 0,1,2,OR 3 FOR AN OPERATION CODE."
160  INPUT C
170  IF C=0 THEN 700
175  IF C=2 THEN 300
180  IF C=3 THEN 400
185  IF C<>1 THEN 150
200  PRINT "INPUT THE NUMBERS TO BE ADDED, ONE AT A TIME."
210  PRINT "WHEN FINISHED TYPE 9E9."
215  LET S=0
220  INPUT S1
230  IF S1>=9E9 THEN 260
240  LET S=S+S1
250  GOTO 220
260  PRINT "THE SUM IS ";S
270  GOTO 150
```

```
300  PRINT "INPUT THE FACTORS TO BE MULTIPLIED, ONE AT A TIME."
310  PRINT "WHEN FINISHED TYPE 9E9."
315  LET M=1
320  INPUT M1
330  IF M1>=9E9 THEN 360
340  LET M=M*M1
350  GOTO 320
360  PRINT "THE PRODUCT IS ";M
370  GOTO 150
400  PRINT "INPUT THE DIVIDEND";
410  INPUT D1
420  PRINT "INPUT THE DIVISOR";
430  INPUT D2
440  PRINT "THE QUOTIENT IS ";D1/D2
450  GOTO 150
700  PRINT "END OF MIDICALC PROGRAM."
710  END
```

2D-2: COMPACT Reduce the size of the above program by using the GOTO K OF or the ON K GOTO statements of BASIC (see page 40). HINT: Try GOTO C+1 OF . . . .

Unit 2E. The MAXICALC Simulator

There are many other "special" operations one could add to a calculator. In this unit we will suggest two of these, and then invite you to build in others.

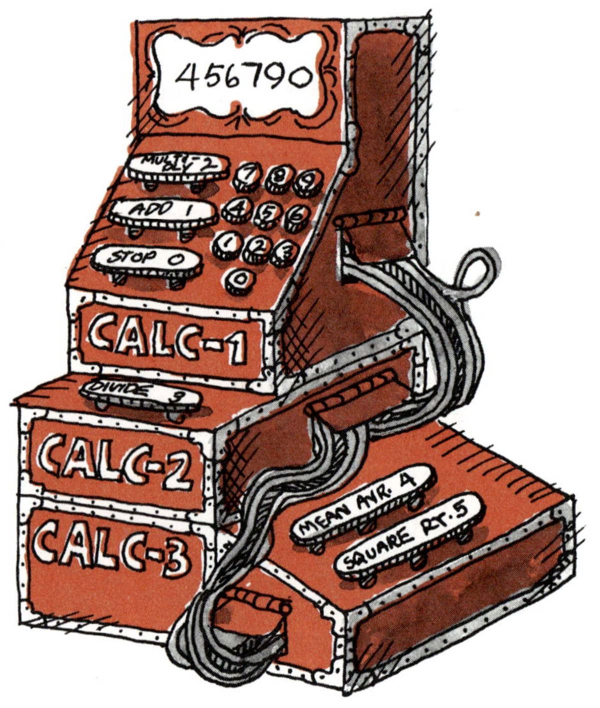

2E-1: CALC-3 Write a program based on CALC-2, adding the two operations "arithmetic averaging" and "taking the square root." Recall that the *average* of n numbers is found by dividing their sum by n. For example, the arithmetic average (*mean*) of the numbers 95, 80, 75, 70 is $(95 + 80 + 75 + 70)/4 = (320)/4 = 80$.

The positive square root of a positive number can be found by using the SQR function (page 20).

Use your MAXICALC simulator program to find the values of:

(a) $\sqrt{4^2 - 4(-35)}$ (b) $\sqrt{\dfrac{2^2 + 5^2 + 4^2}{3}}$

2E-2: SPECIALOP Add other special operations to your calculator simulator. (Example: An operation which takes as input principal amount, annual interest rate, number of years, and number of times compounded per year, and gives as output the total interest accumulated).

SECTION 3 LINEAR EQUATIONS INVOLVING ONE VARIABLE

Checklist of Computing Skills

Previously explained:	PRINT, END, LET, INPUT, IF . . . THEN, STOP, GOTO, FOR, NEXT, ABS, INT, PRINT TAB
Explained in this section:	Subscripted variables, DIM, REM

Checklist of Algebraic Skills

> NOTE: In some algebra books, uniform (or average) velocity is called *rate*, and the formula for uniform motion is given as
>
> distance = rate × time   or $d = rt$.
>
> In the exercises below, V means average velocity.

1. If D = V∗T, and V = 10 kilometers per hour and T = 3 hours, then D = __?__ km.
2. If D = V∗T, and D = 100 kilometers and V = 40 kilometers per hour, then T = __?__ h.
3. If D = V∗T, and D = 850 kilometers and T = 8 hours and 30 minutes, then V = __?__ km/h.
4. If you leave your house at 9:50 A.M. and arrive at the supermarket at 10:00 A.M., and if the supermarket is 1 kilometer from your house, then your average walking velocity is = __?__ kilometers per hour.
5. If 2∗X−5=9, then X = __?__.

When you have finished, check your answers with those at the bottom of the page.

NEW: Subscripted variables; DIM

Up to now the variables allowed in BASIC have been:

(a) Single letters: A,B,X, . . .
(b) Single letters followed by a single digit: A1,B3,X7, . . .

Sometimes it is convenient to be able to use a collection of variables. This can be done by using a *subscripted variable*. ["Subscripted" means "written below", as in $X_1, X_2, X_3, \ldots$. However, computers have to use X(1),X(2),X(3), because of the limitations of terminal keyboards.]

X(K) — K is the subscript. It *must* be in parentheses.

The big advantage of using a subscripted variable is that the *program* can control which variable of the set is to be used, simply by changing K. For

Answers: (1) 30 kilometers (2) 2.5 h (3) 100 km/h (4) 3 km/h (5) 7 example, when K takes on the values 1,2,3, . . . , the subscripted variables selected will be X(1),X(2),X(3), . . . This allows us to store numbers in a lot of variables in one part of a program, and then let the computer get them back out for us in another part (you'll see this in our next example.)

If you want to use subscripts that are greater than 10, you must warn the computer with a DIMension statement. For example,

```
10  DIM X(12)
```

warns the computer that you will need 12 memory locations labeled X(1),X(2), . . . ,X(12). (Some computers print brackets instead of parentheses.)

Study the following program and its RUN. Notice that using the subscripted variable X(K) allows us to first input all the prices and *then* print a neat comparative table.

```
10  DIM X[12]
20  PRINT "WHAT ARE THE PRICES YOU WANT ANALYZED"
30  FOR K=1 TO 12
40  INPUT X[K]
50  NEXT K
60  PRINT
70  PRINT "PRICE","INCREASED 5%","INCREASED 7.5%"
80  FOR K=1 TO 12
90  PRINT X[K],1.05*X[K],1.075*X[K]
100  NEXT K
110  END
RUN


WHAT ARE THE PRICES YOU WANT ANALYZED
?2.98
?159.65
?0.89
?310.57
?3987.98
?16.67
?199.16
?369.78
?545.63
?4.58
?417.38
?59.42

PRICE           INCREASED 5%    INCREASED 7.5%
 2.98            3.129           3.2035
 159.65          167.632         171.624
 .89             .9345           .95675
 310.57          326.099         333.863
 3987.98         4187.38         4287.08
 16.67           17.5035         17.9202
 199.16          209.118         214.097
 369.78          388.269         397.514
 545.63          572.911         586.552
 4.58            4.809           4.9235
 417.38          438.249         448.684
 59.42           62.391          63.8765
```

NEW: REM

REMark statements are used to insert comments into programs. They do *not* print during a RUN—only during a LIST. The following program shows a REMark statement. It also illustrates the use of expressions as subscripts.

```
10  REM PROGRAM TO GENERATE 10 FIBONACCI NUMBERS
20  LET A(1)=1
25  PRINT A(1);
30  LET A(2)=1
35  PRINT A(2);
40  FOR J=3 TO 10
50  LET A(J)=A(J-1)+A(J-2)
55  PRINT A(J);
60  NEXT J
70  END
RUN

 1 1 2 3 5 8 13 21 34 55
```

One more thing—variables with double subscripts like A(K,J) are also allowed. We'll show you a use of these in Section 6.

Unit 3A. Solving Linear Equations

One of the most important skills you will want to acquire early in your study of algebra is the ability to solve equations. For example, you will be interested in solving equations of the form

$$-7x + 18 = 109$$

where x is a member of the set of real numbers. (This is called solving a *linear* equation in *one variable*.)

You have learned how to solve such an equation by first adding (in this case) -18 to both sides (or subtracting 18 from both sides):

$$-7x + 18 + (-18) = 109 + (-18)$$
$$-7x + 18 - 18 = 109 - 18$$
$$-7x = 91$$

Then you multiply both sides by $-1/7$ (or divide both sides by -7):

$$-\frac{1}{7}\cdot(-7x) = -\frac{1}{7}\cdot(91)$$

Thus, $$x = -13.$$

3A-1: LINEAR Let's see how to write a program that will give you (or a friend) practice in solving linear equations. The program is shown on the next page. Instead of explaining how it works with a flowchart, we will give explanatory information next to the appropriate statements. By reading each explanation and then looking at the program, you should get a good idea of how it works.

EXPLANATION	PROGRAM
K will count the number of correct answers; W will count the number of wrong answers.	`5 LET K=0` `7 LET W=0`

Explains the problem.	`10 PRINT "THIS IS A PRACTICE DRILL IN SOLVING";`
	`15 PRINT " THE EQUATION"`
	`20 PRINT "      AX + B = C,"`
	`30 PRINT "WHERE YOU SUPPLY VALUES FOR A,B,C."`
	`35 PRINT`

Allows the student to select real numbers of his choosing for A, B, and C.	`40 PRINT "WHAT VALUE OF A WOULD YOU LIKE";`
	`50 INPUT A`
	`60 PRINT "WHAT VALUE OF B WOULD YOU LIKE";`
	`70 INPUT B`
	`80 PRINT "WHAT VALUE OF C WOULD YOU LIKE";`
	`90 INPUT C`

Repeats the problem with A, B, and C inserted.	`100 PRINT "OK. THE PROBLEM YOU HAVE CHOSEN IS";`
	`105 PRINT " TO FIND X SUCH THAT"`
	`110 PRINT "        ";A;"X + (";B;") = ";C`
	`120 PRINT "WHAT IS X";`

Student gets a chance to answer.	`130 INPUT X`
The computer calculates the correct root.	`140 LET R=(C-B)/A`
Checks student's answer against correct root within ±0.01.	`150 IF ABS(X-R)<.01 THEN 290`

Student with incorrect answer gets explanation of correct method.	`160 PRINT "NO, YOU SHOULD FIRST SUBTRACT ";B;`
	`165 PRINT "FROM BOTH SIDES"`
	`170 PRINT "WHICH GIVES ";A;"X = ";C;" - (";B;`
	`175 PRINT ") = ";C-B`
	`180 PRINT "THEN YOU SHOULD DIVIDE BOTH SIDES";`
	`185 PRINT " BY ";A;", WHICH GIVES"`
	`190 PRINT "X = ";C-B;" /";A;" = ";(C-B)/A`

Increases his "number wrong" counter by 1.	`200 LET W=W+1`
	`205 PRINT`
	`210 PRINT "DO YOU WANT TO TRY ANOTHER?"`
	`220 GOTO 310`

Student with correct answer has the K counter increased by 1.	`290 LET K=K+1` `295 PRINT`

Student gets a chance to try again.	`300 PRINT "THAT'S CORRECT!! WOULD YOU LIKE TO";`
	`305 PRINT " DO ANOTHER?"`
	`310 PRINT "TYPE 1 FOR YES, 2 FOR NO."`
	`320 INPUT A`
	`330 IF A=1 THEN 40`
	`335 PRINT`

Student's grade is printed.	`400 PRINT "YOU HAD ";K;" CORRECT, ";W;" WRONG.";`
	`405 PRINT " YOUR GRADE"`
	`410 PRINT "WAS ";100*K/(K+W);"%"`
	`420 PRINT "THIS IS YOUR ROBOT COACH SIGNING OFF."`
	`500 END`

Here is a sample run of program 3A-1:

```
RUN


THIS IS A PRACTICE DRILL IN SOLVING THE EQUATION
     AX + B = C,
WHERE YOU SUPPLY VALUES FOR A,B,C.

WHAT VALUE OF A WOULD YOU LIKE?3
WHAT VALUE OF B WOULD YOU LIKE?-5
WHAT VALUE OF C WOULD YOU LIKE?4
OK.  THE PROBLEM YOU HAVE CHOSEN IS TO FIND X SUCH THAT
         3X + (-5) =  4
WHAT IS X?3

THAT'S CORRECT!!  WOULD YOU LIKE TO DO ANOTHER?
TYPE 1 FOR YES, 2 FOR NO.
?1
WHAT VALUE OF A WOULD YOU LIKE?5
WHAT VALUE OF B WOULD YOU LIKE?98
WHAT VALUE OF C WOULD YOU LIKE?33
OK.  THE PROBLEM YOU HAVE CHOSEN IS TO FIND X SUCH THAT
         5X + ( 98) =  33
WHAT IS X?-12
NO, YOU SHOULD FIRST SUBTRACT  98 FROM BOTH SIDES
WHICH GIVES  5X =  33 - ( 98) = -65
THEN YOU SHOULD DIVIDE BOTH SIDES BY  5, WHICH GIVES
X = -65 / 5 = -13

DO YOU WANT TO TRY ANOTHER?
TYPE 1 FOR YES, 2 FOR NO.
?1
WHAT VALUE OF A WOULD YOU LIKE?3
WHAT VALUE OF B WOULD YOU LIKE?6
WHAT VALUE OF C WOULD YOU LIKE?9
OK.  THE PROBLEM YOU HAVE CHOSEN IS TO FIND X SUCH THAT
         3X + ( 6) =  9
WHAT IS X?1

THAT'S CORRECT!!  WOULD YOU LIKE TO DO ANOTHER?
TYPE 1 FOR YES, 2 FOR NO.
?2

YOU HAD  2 CORRECT,  1 WRONG.  YOUR GRADE
WAS  66.6667%
THIS IS YOUR ROBOT COACH SIGNING OFF.

END
```

3A-2: XX Modify program 3A-1 so that it presents problems which have X on both sides of the equation.

For example: $3X + 14 = 22X - 7$

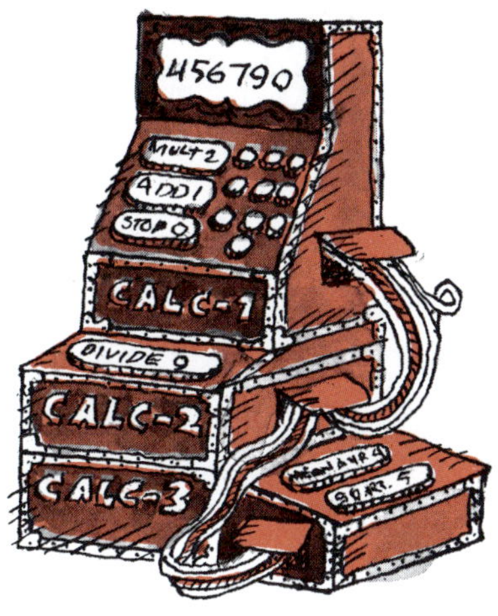

Unit 3B. Solving Linear Equations with MINICALC

It would be nice if a student could practice more difficult problems such as

$$5.6376x + 3.1416 = -18.123$$

when running the "coaching" program 3A-1. This would be possible if he could call on a desk calculator program like those in units 2C, 2D, and 2E.

Here's the plan—when a student is asked for x, instead of answering, suppose he types in 9E9 as a *signal* that he wants a minicalculator. We'll then branch him to line 800:

A new step is added to find out if the user has typed in the *"signal."*

```
 .  (These steps are the same as before.)
 .
135 IF X > 8E9 THEN 800
 .
 .
 .
500 STOP
```

The user can calculate his answer using this program, and *then* answer the original question back at 130.

```
800
 .  (a minicalculator program)
 .
 .
890 PRINT "NOW ANSWER THE QUESTION: WHAT IS X";
899 GOTO 130
900 END
```

Let's illustrate this, using a minicalculator program that can only add or divide:

3B-1: LINEAR-CALC Improve program 3A-1 by *adding* a line 135 that says

```
135  IF X>8E9 THEN 800
```

and *changing* line 500 to STOP, and then *adding* lines 800 to 900:

```
800  PRINT "TO ADD, TYPE 1; TO DIVIDE, TYPE 2; TO LEAVE"
805  PRINT "THE MINICALCULATOR, TYPE 3."
810  INPUT D
815  IF D>2 THEN 890
820  IF D>1 THEN 850
825  PRINT "ADDITION.  TYPE THE NUMBERS YOU WISH TO ADD:"
830  INPUT N1,N2
835  PRINT "THE SUM OF THESE NUMBERS IS ";N1+N2
840  GOTO 800
850  PRINT "TYPE IN THE DIVIDEND:"
855  INPUT D1
860  PRINT "TYPE IN THE DIVISOR:"
870  PRINT "THE QUOTIENT IS ";D1/D2
880  GOTO 800
890  PRINT "NOW ANSWER THE QUESTION:  WHAT IS X";
899  GOTO 130
900  END
```

Here's what a RUN should be like:

```
RUN


THIS IS A PRACTICE DRILL IN SOLVING THE EQUATION
     AX + B = C,
WHERE YOU SUPPLY VALUES FOR A,B,C.

WHAT VALUE OF A WOULD YOU LIKE?5.6376
WHAT VALUE OF B WOULD YOU LIKE?3.1416
WHAT VALUE OF C WOULD YOU LIKE?-18.123
OK.  THE PROBLEM YOU HAVE CHOSEN IS TO FIND X SUCH THAT
         5.6376X + ( 3.1416) = -18.123
WHAT IS X?9E9
TO ADD, TYPE 1; TO DIVIDE, TYPE 2; TO LEAVE
THE MINICALCULATOR, TYPE 3.
?1
ADDITION.  TYPE THE NUMBERS YOU WISH TO ADD:
?-18.123,-3.1416
THE SUM OF THESE NUMBERS IS -21.2646
TO ADD, TYPE 1; TO DIVIDE, TYPE 2; TO LEAVE
THE MINICALCULATOR, TYPE 3.
?2
TYPE IN THE DIVIDEND:
?-21.2646
TYPE IN THE DIVISOR:
?5.6376
THE QUOTIENT IS -3.77192
TO ADD, TYPE 1; TO DIVIDE, TYPE 2; TO LEAVE
THE MINICALCULATOR, TYPE 3.
?3
NOW ANSWER THE QUESTION:  WHAT IS X?-3.77

THAT'S CORRECT!!  WOULD YOU LIKE TO DO ANOTHER?
TYPE 1 FOR YES, 2 FOR NO.
```

Don't forget—program 3B-1 requires *all* the statements from 3A-1, *plus* line 135, and *plus* lines 800 to 899, with line 500 changed to STOP.

Unit 3C. Space-Probe Navigator, Phase I

Units 3C, 3D, and 3E are related to each other. Program 3D-1 is formed by adding lines to program 3C-1, and program 3E-1 can be formed by adding lines to program 3D-1.

3C-1: MOTION The space-probe navigator programs are based on the formula D = V * T which says that the distance traveled by any moving object is found by multiplying its average velocity V by the *elapsed* time, T.

We will assume that our moving object (anything from a bicycle to a space ship) passes known positions, which we call way-STAtions (or STA for short), and that we know the distance between these STAtions.

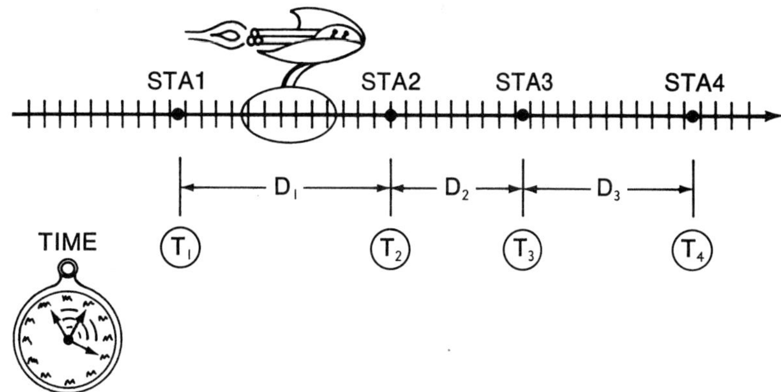

Each time a STA is passed, the time is recorded. (We shall record only the number of minutes and seconds past the hour, since we are assuming that less than an hour is needed for each leg of the journey.)

Calculation of V

We can calculate the average velocity of our ship as soon as it has passed *two* "way-stations." For example:

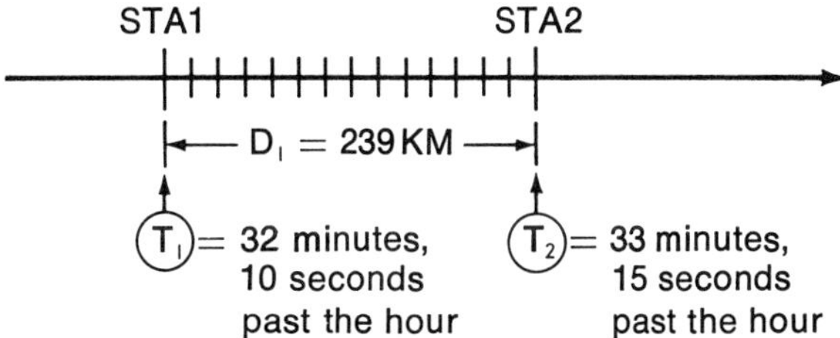

The elapsed time is the difference $T_2 - T_1$. To subtract, we first change T_1 and T_2 into "seconds past the hour":

$$\begin{aligned} T_2 &= 33 \times 60 + 15 = 1{,}995 \text{ seconds} \\ T_1 &= 32 \times 60 + 10 = \underline{1{,}930} \text{ seconds} \\ T_2 - T_1 &= \quad 65 \text{ seconds} \end{aligned}$$

In other words, it takes 65 seconds for our ship to travel 239 km. This gives us:

$$V = \frac{D_1}{T} = \frac{239 \text{ km}}{65 \text{ seconds}}$$

$$\doteq 3.67692 \text{ km per second}$$

If we wish V in kilometers per minute, we would multiply this value by 60, since there are 60 seconds in a minute:

$$V \doteq 3.67692*60 \doteq 220.615 \text{ km per minute}$$

If we wish V in km per hour, we would use another factor of 60 and obtain:

$$V \doteq 3.67692*60*60 \doteq 13236.9 \text{ km per hour}$$

As you can see, there is a lot of work involved in these calculations. Let's write a computer program that does the work for us. Here's a rough flow chart for such a program:

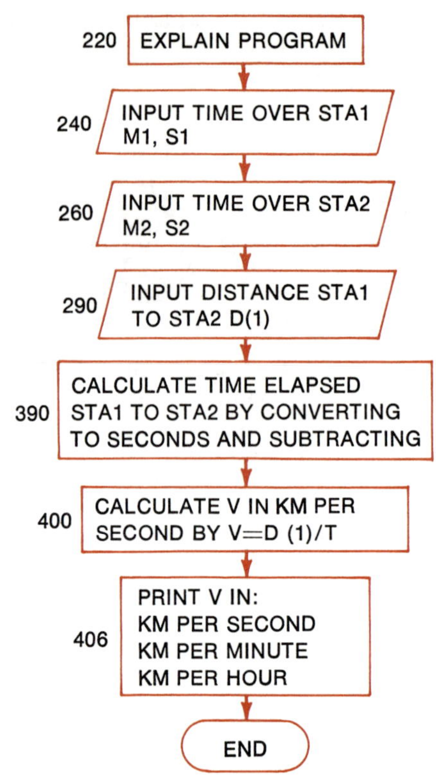


```
220  PRINT "ANALYSIS OF MOTION PROGRAM"
230  PRINT "INPUT TIME OVER STA1 (MINUTES,SECONDS PAST";
235  PRINT " THE HOUR)"
240  INPUT M1,S1
250  PRINT "INPUT TIME OVER STA2 (MINUTES,SECONDS PAST";
255  PRINT " THE HOUR)"
260  INPUT M2,S2
280  PRINT "INPUT DISTANCE FROM STA1 TO STA2 (KM)"
290  INPUT D[1]
390  LET T=(M2*60+S2)-(M1*60+S1)
400  LET V=D[1]/T
402  PRINT "VELOCITY ANALYSIS BASED ON STA1-TO-STA2 LEG"
404  PRINT  "KM PER SEC", "KM PER MIN", "KM PER HR"
406  PRINT V,V*60,V*60*60
630  END
```

Notice that we can convert from KM PER SEC to KM PER MIN and KM PER HOUR right in the print statement.

You'll notice that we used a subscripted variable in our program called D(1). (You'll see why we did this when we get to Unit 3E.) Here's a sample RUN of the program 3C-1.

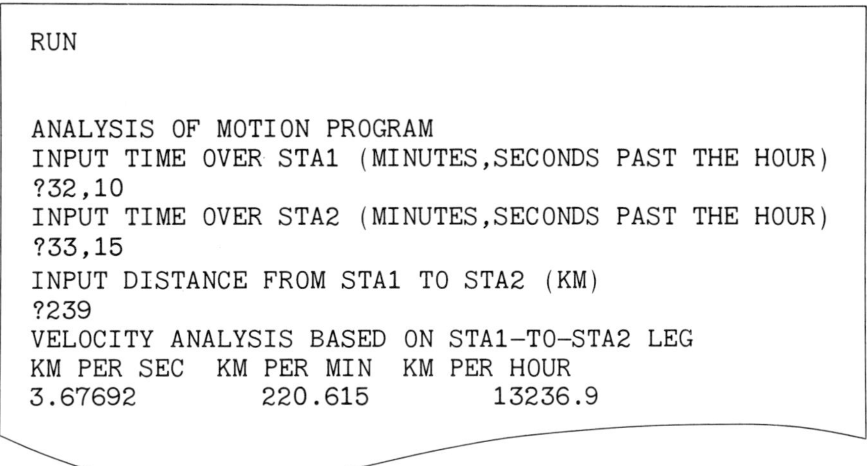

```
RUN


ANALYSIS OF MOTION PROGRAM
INPUT TIME OVER STA1 (MINUTES,SECONDS PAST THE HOUR)
?32,10
INPUT TIME OVER STA2 (MINUTES,SECONDS PAST THE HOUR)
?33,15
INPUT DISTANCE FROM STA1 TO STA2 (KM)
?239
VELOCITY ANALYSIS BASED ON STA1-TO-STA2 LEG
KM PER SEC   KM PER MIN   KM PER HOUR
3.67692        220.615        13236.9
```

Try program 3C-1 for the following three cases:

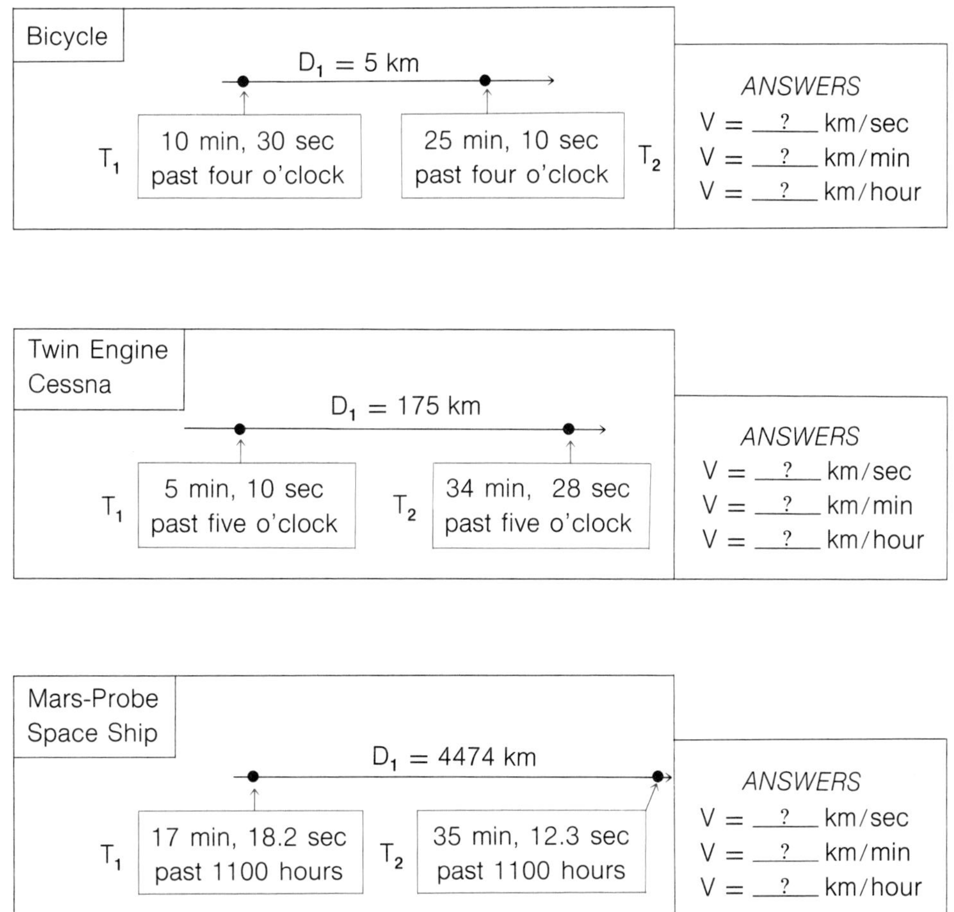

We have assumed that the "hour" in the times given doesn't change because of the short time span involved. But even with short time spans we *could* have the following happen:

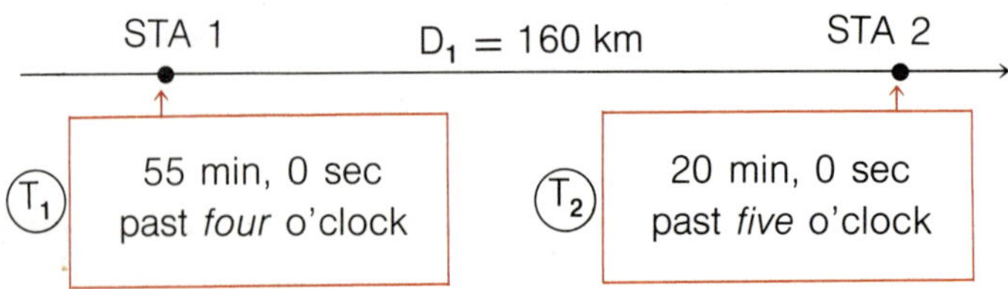

Notice that program 3C-1 won't work now, because line 390 would give us:

$$T = 20 * 60 + 0 - 55 * 60 - 0 = 1200 - 3300$$
$$T = -2100 \text{ seconds}$$

which is wrong!

The right answer is that

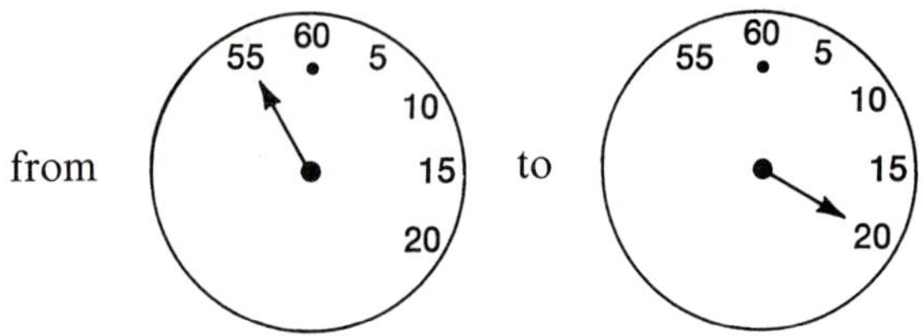

is 25 min. or $T = (25) * (60) = 1{,}500$ seconds.

3C-2: CLOCK Add statements to program 3C-1 so that it automatically detects and handles this unusual case, without your inputting the hour into the program. You can assume that the difference between T_1 and T_2 is less than 60 minutes.

Unit 3D. Space-Probe Navigator, Phase II

In this unit we will add additional steps to program 3C-1 which cause the terminal to give a "picture" of the progress through space of our moving object. Such a picture is referred to as *graphical output*. It has the advantage that someone who is busy (say the captain of our ship) can immediately "see" what is happening. Here's a sample of what we mean:

3D-1: TIME-MAP (STA1 TO STA2)

```
RUN


ANALYSIS OF MOTION PROGRAM
INPUT TIME OVER STA1 (MINUTES,SECONDS PAST THE HOUR)
?32,10
INPUT TIME OVER STA2 (MINUTES,SECONDS PAST THE HOUR)
?33,15
INPUT DISTANCE FROM STA1 TO STA2 (KM)
?239
```

```
VELOCITY ANALYSIS BASED ON STA1-TO-STA2 LEG
KM PER SEC  KM PER MIN  KM PER HOUR
 3.67692     220.615     13236.9

               EACH * = 18.3846 KILOMETERS

CLOCK ETA   GRAPHIC DISTANCE   APPROX. DIST.   STA FIX
 32 : 10        +++++++          0               <---STA 1
 32 : 15           *             18
 32 : 20           *             37
 32 : 25           *             55
 32 : 30           *             74
 32 : 35           *             92
 32 : 40           *             110
 32 : 45           *             129
 32 : 50           *             147
 32 : 55           *             165
 33 : 0            *             184
 33 : 5            *             202
 33 : 10           *             221
 33 : 15        +++++++          239             <---STA 2
END OF OUTPUT
```

We can see at a glance that at 32 minutes and 25 seconds past the hour the ship was about ¼ of the way to STA2.

Here is a listing of program 3D-1. Notice that this is the same as program 3C-1, with step 300 and steps 410 to 620 *added* to give the graphical output.

```
220  PRINT "ANALYSIS OF MOTION PROGRAM"
230  PRINT "INPUT TIME OVER STA1 (MINUTES,SECONDS PAST";
235  PRINT " THE HOUR)"
240  INPUT M1,S1
250  PRINT "INPUT TIME OVER STA2 (MINUTES,SECONDS PAST";
255  PRINT " THE HOUR)"
260  INPUT M2,S2
280  PRINT "INPUT DISTANCE FROM STA1 TO STA2 (KM)"
290  INPUT D[1]
300  LET I=1
390  LET T=(M2*60+S2)-(M1*60+S1)
400  LET V=D[1]/T
402  PRINT "VELOCITY ANALYSIS BASED ON STA1-TO-STA2 LEG"
404  PRINT "KM PER SEC","KM PER MIN","KM PER HOUR"
406  PRINT V,V*60,V*60*60
410  LET N=5
442  PRINT
443  LET U=V*N
445  PRINT TAB(15);"EACH * =";U;" KM"
447  PRINT
450  PRINT "CLOCK ETA";TAB(10);"GRAPHIC DISTANCE";
451  LET E=E1=0
452  PRINT TAB(30);"APPROX. DIST.";TAB(50);"STA FIX"
470  LET K=1
480  PRINT M1;" :";S1;TAB(15);"+++++++";TAB(35);E1;
481  PRINT TAB(50);"<---STA";K
490  IF K>I THEN 620
500  LET T1=D[K]/V
510  LET T2=0
520  LET S1=S1+N
530  LET T2=T2+N
540  IF S1<60 THEN 570
```

```
550  LET M1=M1+1
560  LET S1=S1-60
570  LET E=E+U
572  LET E1=INT(E+.5)
575  IF T2+N>T1 THEN 600
580  PRINT M1;" :";S1;TAB(18);"*";TAB(35);E1
590  GOTO 520
600  LET K=K+1
610  GOTO 480
620  PRINT "END OF OUTPUT"
630  END
```

Here is an explanation of the extra steps in program 3D-1:

Line	Explanation
300	I is the number of STA used.
410	N is the number of seconds we have decided to use between each line of graphic output.
445	U = V*N is the distance traveled in five seconds (N = 5).
451	E is the "accumulated" distance. It is accumulated in step 570. The formula E1 = INT(E + .5) causes the value of E to be "rounded off" to the nearest integer. (If necessary, look back at page 22.)
470	K is the STA number.
480 and 481	Used to print the first and last line of graphical output.
490	Stops the program when we reach STA2.
500	T1 is the number of seconds from STA1 to STA2, since D(K) = D(1). When you get to Unit 3E, you'll see why we introduced K.
510 to 560	These lines control the ETA (estimated time of arrival). They cause the time to print every N seconds, and they make the minutes (M1) increase by 1 every time the seconds reach 60.

3D-2: CONDENSE It could turn out that printing a line of graphical output every N=5 seconds is too often, and that this uses up too much paper. Modify program 3D-1 so that a limit of 15 lines of graphical output is printed between stations.

Your program should change N to the smallest multiple of 5 (five seconds, ten seconds, fifteen seconds, and so on) that still permits up to 15 lines (<=14) of output to be printed between any two STAs.

HINT: Would the following help?

```
410  LET N=5
420  IF T/N <=14 THEN 442
430  LET N=N+5
440  GOTO 420
442  PRINT
```

Unit 3E. Space-Probe Navigator, Phase III

The real value of the information that we have calculated so far is that it can be used to *predict* the ETA (estimated time of arrival) over all the remaining STA points in a journey at a constant velocity. Here's how it works:

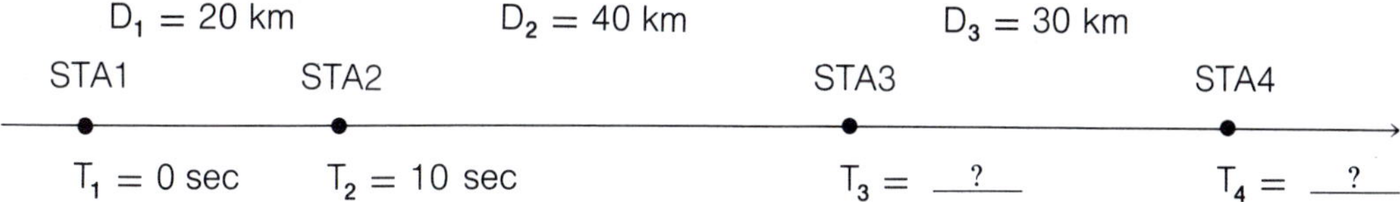

1. We calculate V from D_1 and $T_2 - T_1$ (see program 2C-1). $V = 20/(10 - 0) = 20/10 = 2$ km per second
2. Now that we know V, we calculate T_3. $(T_3 - T_2) = D_2/V = 40/2 = 20$ seconds. But we know that $T_2 = 10$ seconds. $T_3 - 10 = 20$ or $T_3 = 30$ seconds from start
3. Now we repeat the process for the next leg. $(T_4 - T_3) = D_3/V = 30/2 = 15$ seconds. But we now know T_3 from step 2: $T_3 = 30$ $T_4 - 30 = 15$ or $T_4 = 45$ seconds from start

3E-1: PREDICT-MAP Write a program which puts these ideas into graphical form, giving a map of the entire journey for any number of STATIONS. The first part of the RUN should request input data. Our example uses the two "times" (T_1 at STA1 and T_2 at STA2) from program 3D-1, and four distances:

D1	(STA1 to STA2)	239 km
D2	(STA2 to STA3)	315 km
D3	(STA3 to STA4)	189 km
D4	(STA4 to STA5)	227 km

Your output should look something like that shown below. Notice that the rounding of the amount represented by the *'s makes the computed "APPROX. DIST." differ a little from the actual distance. That is, the distance from STA1 to STA3 is actually 239 + 315 or 554 km instead of the 552 given by this program.

```
RUN


ANALYSIS OF MOTION PROGRAM
INPUT TIME OVER STA1 (MINUTES,SECONDS PAST THE HOUR)
?32,10
INPUT TIME OVER STA2 (MINUTES,SECONDS PAST THE HOUR)
?33,15
INPUT NUMBER OF STATIONS
?5
INPUT D 1?239
INPUT D 2?315
INPUT D 3?189
INPUT D 4?227
```

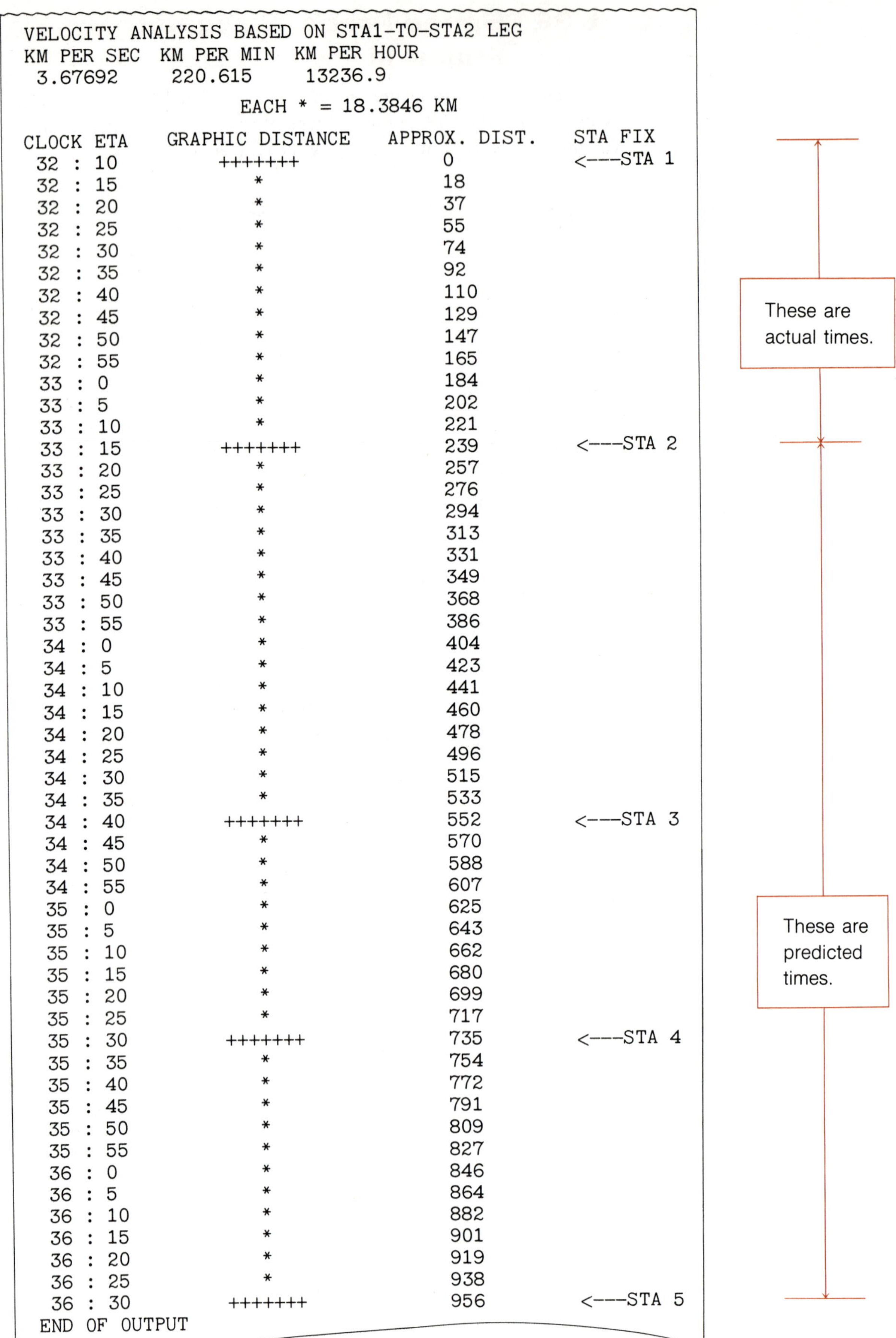

```
VELOCITY ANALYSIS BASED ON STA1-TO-STA2 LEG
KM PER SEC  KM PER MIN  KM PER HOUR
 3.67692      220.615     13236.9

                     EACH * = 18.3846 KM

CLOCK ETA    GRAPHIC DISTANCE   APPROX. DIST.    STA FIX
 32 : 10         +++++++            0            <---STA 1
 32 : 15            *               18
 32 : 20            *               37
 32 : 25            *               55
 32 : 30            *               74
 32 : 35            *               92
 32 : 40            *               110
 32 : 45            *               129
 32 : 50            *               147
 32 : 55            *               165
 33 : 0             *               184
 33 : 5             *               202
 33 : 10            *               221
 33 : 15         +++++++            239          <---STA 2
 33 : 20            *               257
 33 : 25            *               276
 33 : 30            *               294
 33 : 35            *               313
 33 : 40            *               331
 33 : 45            *               349
 33 : 50            *               368
 33 : 55            *               386
 34 : 0             *               404
 34 : 5             *               423
 34 : 10            *               441
 34 : 15            *               460
 34 : 20            *               478
 34 : 25            *               496
 34 : 30            *               515
 34 : 35            *               533
 34 : 40         +++++++            552          <---STA 3
 34 : 45            *               570
 34 : 50            *               588
 34 : 55            *               607
 35 : 0             *               625
 35 : 5             *               643
 35 : 10            *               662
 35 : 15            *               680
 35 : 20            *               699
 35 : 25            *               717
 35 : 30         +++++++            735          <---STA 4
 35 : 35            *               754
 35 : 40            *               772
 35 : 45            *               791
 35 : 50            *               809
 35 : 55            *               827
 36 : 0             *               846
 36 : 5             *               864
 36 : 10            *               882
 36 : 15            *               901
 36 : 20            *               919
 36 : 25            *               938
 36 : 30         +++++++            956          <---STA 5
 END OF OUTPUT
```

SECTION 4 USING INEQUALITIES

Checklist of Computing Skills

Previously explained:	PRINT, END, LET, INPUT, IF . . . THEN, STOP, GOTO, FOR, NEXT, ABS, INT, PRINT TAB, subscripted variables—single subscripts
Explained in this section:	RND, READ, DATA
Require extra reading:	File Commands (Used in program 4E-3)

Checklist of Algebraic Skills

1. Which of the numbers listed below is *not* in the solution set of

   $-2.7 \leq x \leq 7.1$?

   -2.6, 7.1, -2.7, -5.0 __?__

2. On a number line, graph the solution set of:

   (a) $-7 < x < -5$ (c) $0 \leq x \leq 3$

   (b) $-4 \leq x < -2$ (d) $5 < x \leq 7$

The solution sets in Exercise 2 are sometimes called **intervals.** We may denote a (closed) interval as $\{x: a \leq x \leq b\}$. The length of the interval $\{x: a \leq x \leq b\}$ is $b - a$.

3. Express the solution set of $|x - 5| \leq 3$ as an interval. __?__
4. What is the length of the interval found in Exercise 3? __?__
5. What is the length of the interval $-11.1 \leq y \leq -6.22$? __?__
6. What is the length of the interval $-2.1 \leq w \leq 0.9$? __?__

When you have finished, check your answers with those printed at the bottom of the page.

NEW: RND (Random number generator)

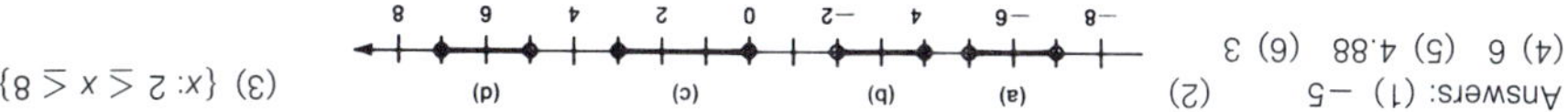

Answers: (1) -5 (2) (4) 6 (5) 4.88 (6) 3 (3) $\{x: 2 \leq x \leq 8\}$

You can think of a random number as a "surprise" number that is chosen by the computer and stored in one of its memory locations. In BASIC, you tell the computer to do this with a statement of the form:

```
40  LET X = RND(1)
```

The surprise, or random, number which is stored in location X, is usually chosen to be a decimal number greater than zero, but less than 1. In the language of algebra, the variable X will be a "real" number described by: $0 < X < 1$. However, with a little ingenuity we can convert X into an integer which falls in any desired interval. This will be explained on pages 74–75.

The general form of the function is RND(Z). On some computers, the value of Z is not important; on other computers, it makes a difference. To see how your computer works, run the following program *twice*.

```
10  FOR K=1 TO 5
20  PRINT RND(1)
30  NEXT K
40  END
```

Here's the result of the program above on two different computer systems, which we'll call A and B.

COMPUTER A

```
RUN

 .529432        .225555        .329078        .306689        .537845

END
RUN

 .696617        .940248        8.94691E-02    9.68938E-02    .64532

END
```

COMPUTER B

```
RUN

 .529432        .225555        .329078        .306689        .537845

END
RUN

 .529432        .225555        .329078        .306689        .537845

END
```

Computer A produced a completely *different* set of random numbers on each RUN. For the applications in this book, this is preferred. If your computer acted like computer A you're all set!

If your computer acted like computer B, there are three things you can try to make it act like computer A, producing a real "surprise" on every RUN.

[1] On some systems, you add a "RANDOMIZE" statement at the beginning of the program. Try this experiment:

```
5  RANDOMIZE
10  FOR K=1 TO 5
20  PRINT RND(1),
30  NEXT K
40  END
```

[2] On other systems, the way to get different random numbers on every RUN is to change statement 5 to read:

5 LET X=RND (−1)

The rest of the program stays the same.

[3] If neither of the above work, there is a somewhat clumsy way of making each RUN be "almost" a surprise. It takes five extra statements as follows:

```
5  PRINT "TYPE THE TIME (TIME 9:45 AS 9,45)";
6  INPUT H,M
7  FOR J=1 TO H*M
8  LET X=RND(1)
9  NEXT J
10  FOR K=1 TO 5
20  PRINT RND(1),
30  NEXT K
40  END
```

If a user typed in 2, 6 to indicate that the time "happened" to be 2:06, lines 7, 8, and 9 would force the computer to run down its list of random numbers to the 12th one before printing anything in line 20. If on a second RUN, the time happened to be 3:20, a different number in the list would be used as the starting point.

One last thing—if your computer acts like A, and you *want* it to act like B (to check a program, for example), try experiment [2]. This technique works in reverse on some computers!

Now let's look at a program that uses RND. We'll write a computer program that *simulates* the tossing of a coin eight times. We'll assume that the random numbers are evenly distributed between 0 and 1. Since there are two possible results of a coin toss (HEAD or TAIL), let's decide that if $R<.5$, it represents a HEAD, and if $R\geq.5$, it represents a TAIL (we could just as well reverse this choice).

```
5   LET X=RND(-1)
10  LET H=0
20  FOR I=1 TO 8
30  LET R=RND(1)
40  IF R<.5 THEN 70
50  PRINT " TAILS"
60  GOTO 90
70  LET H=H+1
80  PRINT " HEADS"
90  NEXT I
100  PRINT "NUMBER OF HEADS =";H
110  END
RUN


 TAILS
HEADS
HEADS
HEADS
 TAILS
 TAILS
 TAILS
HEADS
NUMBER OF HEADS = 4

END
```

To get different tosses on different RUNS, your computer may require that you omit this step, or use

```
5  RANDOMIZE
```

instead.

Just as if you tossed a real coin, the order of HEADS and TAILS is random. If you let the program run for a long time, however, say for I = 1 TO 100, it is highly probable that there will be close to 50 heads and 50 tails.

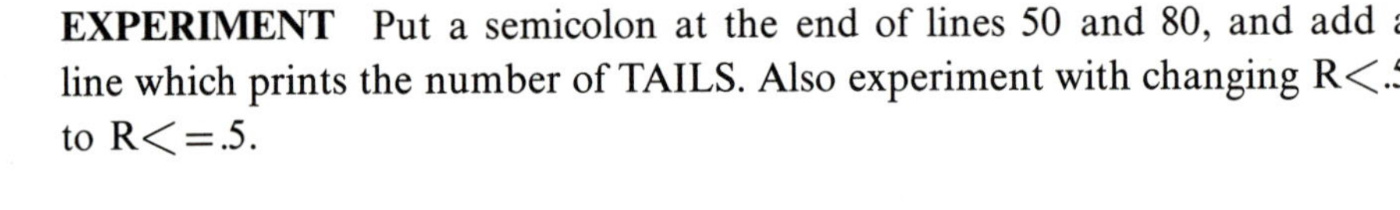

EXPERIMENT Put a semicolon at the end of lines 50 and 80, and add a line which prints the number of TAILS. Also experiment with changing R<.5 to R<=.5.

RND(1) generates decimals between 0 and 1. Frequently, though, we need integers between two other numbers. For instance, to simulate rolling a die, we might want to generate random integers from 1 to 6 (1, 2, 3, 4, 5, 6). Here's how we can make this happen:

RND(1)	gives numbers between 0 and 1 (not including 0 and 1).
6*RND(1)	gives numbers between 0 and 6 (but not including 0 and 6).
INT(6*RND(1))	gives integers 0, 1, 2, 3, 4, 5.
INT(6*RND(1)+1)	gives integers 1, 2, 3, 4, 5, 6, which is what we want.

For example, here is a simulation of the tossing of a single die 25 times:

```
5   (Insert a randomize statement as on page 73 if necessary.)
10  FOR I=1 TO 25
20  PRINT INT(6*RND(1)+1);
30  NEXT I
40  END
RUN

 6 3 4 2 3 3 6 5 4 2 3 4 6 3 5 3 2 2 1 4 2 5 6 4 6
```

In general, INT((S−R+1)*RND(1)+R) gives integers from R to S inclusive. We can say that this formula gives us integers I such that $R \leq I \leq S$.

If R=1 and S=6, you find INT(6*RND(1)+1) as above.

Unit 4A. Random Practice

4A-1: RAND-DIG Write a program that will print out 25 randomly selected single-digit numbers. HINT: Use the formula INT((S−R+1)(* RND(1) + R) with R = 0 and S = 9.

4A-2: RAND-INT Write a program that prints 30 randomly selected integers I where $-50 \leq I \leq +50$. HINT: Let R = −50 and S = 50.

4A-3: RAND-PRICE Print 72 randomly selected prices from $0.99 to $49.99. *Partial Solution:* Let R = 99 and S = 4999. Then S − R + 1 = 4901.

```
10  FOR I = 1 TO 72
20  PRINT "$";(INT(4901 * RND(1) + 99))/100,
```

4A-4: RAND-ADD Here is a program that will give practice in adding pairs of randomly selected two-digit numbers. Compare lines 50 and 60 with the formula at the top of the page. What are the values of R and S here? Try running the program for N = 5.

```
5   (Insert a randomize statement if necessary.)
10  PRINT "DRILL ON CALCULATING A+B"
20  PRINT "HOW MANY PROBLEMS DO YOU WANT";
30  INPUT N
40  FOR I=1 TO N
50  LET A=INT(40*RND(1)+10)
60  LET B=INT(40*RND(1)+10)
70  PRINT "WHAT IS THE VALUE OF";
80  PRINT A;" +";B;
90  INPUT S
100  IF S=A+B THEN 130
110  PRINT "WRONG! THE ANSWER IS";A+B;"."
120  GOTO 140
130  PRINT "RIGHT!"
140  PRINT
150  NEXT I
160  END
```

4A-5: RAND-MULTADD Rewrite program 3A-1 so that the values of A are randomly selected from {1, 2, . . . , 9} and the values of B and C are randomly selected from {10, 11, . . . , 20}, and both multiplication and addition problems are given.

Unit 4B. Random Practice (continued)

4B-1: RAND-LINEAR Change the coaching program 3A-1 (page 59) so that the computer selects A, B, and C with:

A chosen from the set $\{2, 3, 4, 5, 6\}$
B chosen from the set $\{-3, -2, -1, 0, 1, 2, 3\}$
C chosen from the set $\{-20, -10, 0, +10, +20\}$.

HINT: (1) Delete lines 30, 40, 50, 60, 70, 80, 90, and 100. (2) Add the lines:

```
5   (Insert if necessary)
30   PRINT "WHERE THE COMPUTER SELECTS A, B, AND C"
40   LET A=INT(5*RND(1) + 2)
50   LET B=INT(7*RND(1)-3)
60   LET C=10*INT(5*RND(1)-2)
```

4B-2: RAND-XX Convert program 3A-2 so that the random number generator is used to generate the data for the problems.

4B-3: RAND-LINEAR-CALC Convert program 3B-1 so that the random number generator is used to generate data for the problems.

Unit 4C. Simulations and Games

On pages 73–74, you saw how the computer could be programmed to simulate the tossing of a coin, and the tossing of a die. Random numbers can be used in a variety of simulations and games.

4C-1: DICE Write a program that simulates the throwing of two dice. A RUN might look like this:

```
RUN

FIRST DIE        SECOND DIE        TOTAL
 6                2                 8
 1                4                 5
 5                3                 8
 1                2                 3
 6                5                 11
 4                1                 5
 4                6                 10
```

4C-2: GUESS Write a program for a game in which two players guess which number between 1 and 100 the computer has randomly picked. The program should give 10 points to the player who was closer. Assume that the players will choose different numbers. End the game when either player reaches a score of 50. A RUN might look like this:

```
RUN


EACH PLAYER SHOULD CHOOSE AN INTEGER BETWEEN 1 AND 100.

PLAYER 1?24
PLAYER 2?56
THE COMPUTER HAS CHOSEN 15.
PLAYER 1 WAS CLOSER.
PLAYER 1 HAS 10 POINTS.    PLAYER 2 HAS 0 POINTS.

PLAYER 1?50
PLAYER 2?56
THE COMPUTER HAS CHOSEN 53.
TIE--NO CHANGE IN SCORE.

PLAYER 1?
```

```
PLAYER 1?82
PLAYER 2?75
THE COMPUTER HAS CHOSEN 78.
PLAYER 2 WAS CLOSER.
PLAYER 1 HAS 30 POINTS.    PLAYER 2 HAS 50 POINTS.

PLAYER 2 WINS.
```

HINT: Use the absolute value of the difference between the computer's number and a player's number when you are testing for "closeness," since, for example, 50 and 56 are equally "close" to 53.

4C-3: OUTGUESS Write a program for a game in which a person tries to outguess the computer. Here's the way it goes:

1. The computer will secretly select a number.
2. The computer will tell the player in what *interval* the number lies, and the *tolerance* allowed.
3. The player is then asked to guess the computer's number.
4. SCORING:
   (a) If the absolute value of the difference between the player's guess and the computer's number is less than the allowed tolerance, you get 5 points. If not, the computer gets 5 points.

(b) If the player inadvertently guesses outside the given interval, the player loses 5 points.

5. Allow 6 plays for the game.

The tolerance is set in the program as a percent of the length of the interval chosen by the computer. By making the tolerance very small, we can "load" the game in favor of the computer.

Here is a sample RUN of a program in which the tolerance is set at 20% of the length of the interval in which the mystery number lies.

```
RUN

THE COMPUTER HAS CHOSEN A REAL NUMBER C WHERE:
      -8.9 <= C <= .9
WHAT IS YOUR GUESS FOR C?-4
C WAS -5.7.
THE DISTANCE BETWEEN YOUR GUESS AND C IS 1.7.
TOLERANCE = ( .9 - (-8.9)) * .2 = 9.8 * .2 = 1.96
YOU WIN 5 POINTS.                     SCORE*** COMPUTER: 0   YOU: 5
```

```
THE COMPUTER HAS CHOSEN A REAL NUMBER C WHERE:
      -.2 <= C <= 10.7
WHAT IS YOUR GUESS FOR C?5
C WAS 8.4.
THE DISTANCE BETWEEN YOUR GUESS AND C IS 3.4.
TOLERANCE = ( 10.7 - (-.2)) * .2 = 10.9 * .2 = 2.18
I WIN 5 POINTS.                       SCORE***  COMPUTER: 10   YOU: 20


END OF GAME.  YOU WIN.
```

We shall refer to ABS(C − C1) as the *distance between* C and C1 (or between C1 and C).

Let's examine this first play. Here's a picture of what happened on the number line.

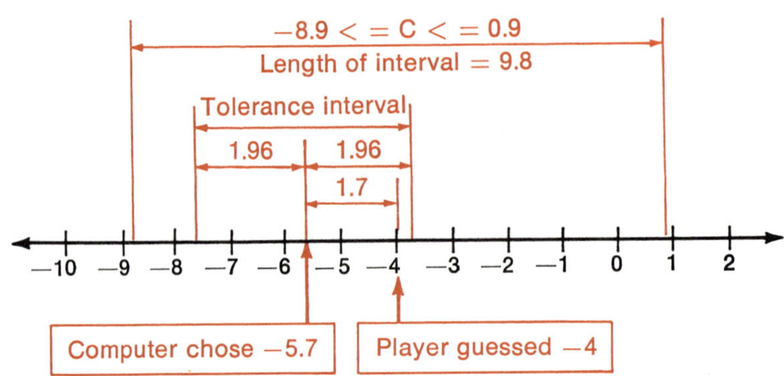

Since ABS(−5.7−(−4)) = 1.7 and 1.7 < 1.96, the player wins 5 points. Here is a listing of program 4A-3 with some comments.

```
10 LET T=.2
20 LET Y=0
30 LET M=0
40 FOR N=1 TO 6
50 LET T1=INT(199*RND(1)-99)/10
60 LET T2=INT(199*RND(1)-99)/10
70 IF T1<T2 THEN 110
80 LET A=T2
90 LET B=T1+1
100 GOTO 130
110 LET A=T1
120 LET B=T2+1
130 LET I=B-A
140 LET C=A+I*RND(1)
150 LET C=INT(10*C)/10
160 PRINT "THE COMPUTER HAS CHOSEN";
170 PRINT "A REAL NUMBER C WHERE:"
180 PRINT "        ";A;" <= C <= ";B
190 PRINT "WHAT IS YOUR GUESS FOR C";
200 INPUT C1
210 IF C1<A THEN 370
220 IF C1>B THEN 370
230 PRINT "C WAS ";C;"."
240 LET D=ABS(C-C1)
250 PRINT "THE DISTANCE BETWEEN YOUR ";
260 PRINT "GUESS AND C IS";D;"."
270 LET L=I*T
280 PRINT "TOLERANCE = (";B;" - (";A;")) *";
290 PRINT T;" =";I;" *";T;" =";L
300 IF D>L THEN 340
310 LET Y=Y+5
320 PRINT "YOU WIN 5 POINTS.";
330 GOTO 390
340 LET M=M+5
350 PRINT "I WIN 5 POINTS.";
360 GOTO 390
370 PRINT "NOT IN INTERVAL--YOU LOSE 5 POINTS.";
380 LET Y=Y-5
390 PRINT TAB(38);"SCORE***  COMPUTER:";M;"   YOU:";Y
400 PRINT
410 NEXT N
420 PRINT
430 PRINT "END OF GAME.";
440 IF Y>M THEN 500
450 IF Y=M THEN 480
460 PRINT "  I WIN."
470 STOP
480 PRINT "  TIE GAME"
490 STOP
500 PRINT "  YOU WIN."
510 END
```

Tolerance is to be 20% of the length of the interval. (line 10)

INT (199*RND(1)−99) gives integers from −99 to 99. Dividing by 10 gives numbers from −9.9 to 9.9 by tenths. (lines 50–60)

Line 70 determines which of T1 and T2 is the smaller. This is renamed A in line 80 or line 110.

These steps insure that the length of the interval will be at least 1. (lines 90 and 120)

Line 140 gives C a value between A and B. Line 150 truncates the value to tenths.

4C-4: TOLERANCE The tolerance in the game in program 4C-3 was set at 20%, or 0.2, of the length of the interval. Change the game so that you can INPUT what tolerance rate (T) you want (from 0.50 to 0.10), BUT your score for a right guess is to be 2/T points!

If you choose 0.10 for T, a right answer gives you 20 points (but you have to be very close).	GO-FOR-BROKE
If you choose 0.40 for T, a right answer gives you 5 points (but you don't have to be very close).	CONSERVATIVE

What's the best strategy?

```
RUN


WHAT TOLERANCE RATE DO YOU WISH?.1
THE COMPUTER HAS CHOSEN A REAL NUMBER C WHERE:
       -1 <= C <=  1.6
WHAT IS YOUR GUESS FOR C?-.5
C WAS -.2.
THE DISTANCE BETWEEN YOUR GUESS AND C IS .3.
TOLERANCE = ( 1.6 - (-1)) * .1 = 2.6 * .1 = .26
I WIN 5 POINTS.                          SCORE***   COMPUTER: 5   YOU: 0

THE COMPUTER HAS CHOSEN A REAL NUMBER C WHERE:
       -4.6 <= C <=  8.3
WHAT IS YOUR GUESS FOR C?6.4
C WAS -4.3.
THE DISTANCE BETWEEN YOUR GUESS AND C IS 10.7.
TOLERANCE = ( 8.3 - (-4.6)) * .1 = 12.9 * .1 = 1.29
I WIN 5 POINTS.                          SCORE***   COMPUTER: 10   YOU: 0

THE COMPUTER HAS CHOSEN A REAL NUMBER C WHERE:
        2.5 <= C <=  5.1
WHAT IS YOUR GUESS FOR C?3.7
C WAS  3.5.
THE DISTANCE BETWEEN YOUR GUESS AND C IS .2.
TOLERANCE = ( 5.1 - ( 2.5)) * .1 = 2.6 * .1 = .26
YOU WIN 2 / .1 = 20 POINTS.              SCORE***   COMPUTER: 10   YOU: 20
```

QUESTION Can you develop a sure-fire "system" for beating the computer in the above game?

4C-5: VICTIM Rewrite the program of 4C-3 so that the tolerance is not controlled by the player. When the game starts, the tolerance should be high to gain the player's confidence. Ask the player to make bets on each run. When the bets get sufficiently high, the program should change the tolerance to a low value so that the machine wipes out its victim. (The scoring should be honest. All you change is the chance that the player has of winning.)

Unit 4D. The Magic Bus Schedule

In this unit we will teach the computer to "read" a bus timetable and tell us when to expect the next bus. Many modern applications of computers (for example airplane or hotel reservation systems) are based on the kind of logic used in this program.

First, we need some additional BASIC statements.

NEW: READ, DATA

These statements are used for storing data (given numbers) within programs. The READ statement contains the variable(s) in which the numbers in the DATA statement are to be stored. The DATA statements can be placed anywhere in the program, but the numbers must be recorded in the order in which they will be called for by the READ statements. A "pointer" moves from each DATA value to the next as the program is executed.

Let's study this feature of BASIC with an example. Here's a program that quizzes you on inches and centimeters.

```
10  PRINT "QUIZ ON THE NUMBER OF CENTIMETERS ";
20  PRINT "IN A NUMBER OF INCHES"
30  PRINT
40  READ I
50  PRINT "HOW MANY CENTIMETERS ARE THERE IN";I;" INCHES";
60  INPUT C
70  IF C=2.54*I THEN 100
80  PRINT "NO.  THE ANSWER IS";2.54*I;" CENTIMETERS."
90  GOTO 30
100  PRINT "CORRECT"
110  GOTO 30
120  DATA 10,20,30
130  DATA 15
140  END
RUN
```

	QUIZ ON THE NUMBER OF CENTIMETERS IN A NUMBER OF INCHES
POINTER IS AT 10 →	HOW MANY CENTIMETERS ARE THERE IN 10 INCHES?25.4 CORRECT
AT 20 →	HOW MANY CENTIMETERS ARE THERE IN 20 INCHES?50.8 CORRECT
AT 30 →	HOW MANY CENTIMETERS ARE THERE IN 30 INCHES?76.2 CORRECT
AT 15 →	HOW MANY CENTIMETERS ARE THERE IN 15 INCHES?38.1 CORRECT
POINTER CAN'T FIND ANY MORE DATA →	OUT OF DATA IN LINE 40

There can be as many DATA statements as is convenient. We can equally well put all the DATA on one line as

```
120  DATA 10,20,30,15
```

or on several lines as:

```
120  DATA 10,20
125  DATA 30
130  DATA 15
```

You can think of the numbers in these statements as being on one continuous list, with a pointer set at the top of the list each time the program is RUN.

10 ← POINTER 20 30 15	In the above program each time statement 40 (READ I) is executed, the number opposite the pointer is stored in the location I,
10 20 ← POINTER 30 15	*and* then the pointer is moved down to the next number on the list.

This process continues until the computer runs out of data, then it prints a message and stops.

Instead of just letting a program run out of data, we can use a special data item (such as a negative number) to signal the end of data.

One other thing you should know: if a READ statement has more than one variable (for example, READ X,Y,Z) it moves the pointer down the list as many positions as needed to get data for each variable—in this case 3.

Here's an example which reads 2 pieces of data at a time, and which uses -1 to signal the end of data. Check it through carefully to make sure that you understand what happens.

```
10  PRINT "LIST","% DISCOUNT","NET"
20  READ L,D
30  IF L<0 THEN 70
40  PRINT L,D,L-L*D/100
50  GOTO 20
60  DATA 55.89,15,106.5,25,33.98,20,-1,0
70  PRINT"FINISHED"
80  END
RUN


LIST            % DISCOUNT      NET
 55.89           15              47.5065
 106.5           25              79.875
 33.98           20              27.184

FINISHED
```

Two values are needed because line 20 reads values in pairs.

EXERCISE Modify and run the above program so that the net prices are rounded off to the nearest penny.

Now let's return to our bus timetable program.

4D-1: ONE-BUS-ROUTE Write a program which requests you to INPUT the time. The program should then check the timetable for that route and print out the scheduled arrival times of the next three buses and how long before each will arrive.

Here is a sample RUN of such a program.

```
RUN


THIS PROGRAM IS FOR BUS ROUTE NO. 16.
IT IS VALID FOR BUS TIMES FROM 1 P.M. TO MIDNIGHT.
IF YOU TYPE IN THE TIME, IT WILL TELL YOU
WHEN THE NEXT THREE BUSES WILL LEAVE.

WHAT TIME IS IT (TYPE IN HOUR,MINUTES)?1,45

THE 1 : 50 BUS WILL LEAVE IN 0 HOURS AND 5 MINUTES.
THE 2 : 45 BUS WILL LEAVE IN 1 HOURS AND 0 MINUTES.
THE 3 : 55 BUS WILL LEAVE IN 2 HOURS AND 10 MINUTES.
END OF PROGRAM
```

Let's try it again with a different time to see what happens when the computer runs out of data.

```
RUN


THIS PROGRAM IS FOR BUS ROUTE NO. 16.
IT IS VALID FOR BUS TIMES FROM 1 P.M. TO MIDNIGHT.
IF YOU TYPE IN THE TIME, IT WILL TELL YOU
WHEN THE NEXT THREE BUSES WILL LEAVE.

WHAT TIME IS IT (TYPE IN HOUR,MINUTES)?11,10

THE 11 : 30 BUS WILL LEAVE IN 0 HOURS AND 20 MINUTES.
NO MORE BUSES FOR THE DAY.
END OF PROGRAM
```

Now look at the flow chart on page 84.

Here is a flow chart for program 4D-1. The numbers beside the boxes are the numbers of corresponding lines in the program.

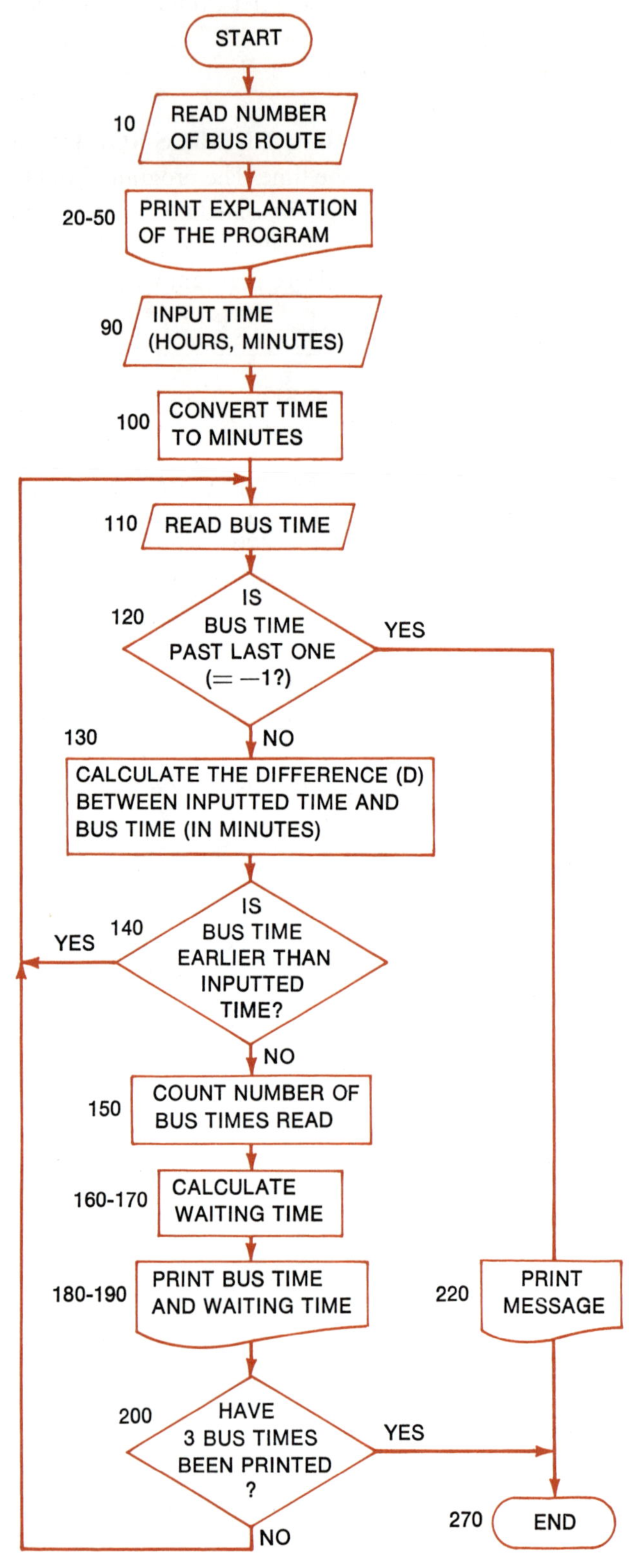

Here is a listing of a program based on the preceding flow chart.

```
10 READ B
20 PRINT "THIS PROGRAM IS FOR BUS ROUTE NO.";B;"."
30 PRINT "IT IS VALID FOR BUS TIMES FROM 1 P.M. TO MIDNIGHT."
40 PRINT "IF YOU TYPE IN THE TIME, IT WILL TELL YOU"
50 PRINT "WHEN THE NEXT THREE BUSES WILL LEAVE."
60 PRINT
70 LET I=0
80 PRINT "WHAT TIME IS IT (TYPE IN HOUR,MINUTES)";
90 INPUT H,M
95 PRINT
100 LET A=60*H+M
110 READ H1,M1
120 IF H1=-1 THEN 220
130 LET D=60*H1+M1-A
140 IF D<0 THEN 110
150 LET I=I+1
160 LET H2=INT(D/60)
170 LET M2=D-H2*60
180 PRINT "THE";H1;" :";M1;" BUS WILL LEAVE IN";
190 PRINT H2;" HOURS AND";M2;" MINUTES."
200 IF I >= 3 THEN 260
210 GOTO 110
220 PRINT "NO MORE BUSES FOR THE DAY."
230 DATA 16
240 DATA 1,5,1,50,2,45,3,55,5,0,6,20,8,5
250 DATA 9,30,10,14,10,55,11,30,-1,-1
260 PRINT "END OF PROGRAM"
270 END
```

We have to have two values here because line 110 is reading values in pairs.

4D-2: MULTI-ROUTE Can you write a program that allows the user to specify which one of three bus routes he wishes to take, and gives him the corresponding times?

4D-3: MIDNIGHT-BUS Can you remove the limitation of allowing only times from 1 P.M. to midnight?

Unit 4E. The Robot Car Salesman

Can a computer help you find the car of your dreams? Probably not. But it's fun to try, which is what the programs in this unit do.

4E-1: HONEST-HORACE The idea is to store characteristics of various cars as DATA in a program. The person who wishes to buy a car then inputs the characteristics he (or she) prefers. The computer searches its DATA to see how closely the cars on file match these preferences. There is no one formula for calculating closeness. We will invent our own and suggest others you may wish to try.

The first thing to do is to put down some characteristics of our imaginary cars and assign them numbers. Here are six we can use for a start. You may wish to change this list later on.

MAKE	ENGINE	COLOR
1. A Motors	1. 4 cylinders	1. White
2. B Motors	2. 6 cylinders	2. Bright
3. C Motors	3. Diesel	3. Dark
4. D Motors	4. Rotary	4. Special finish
5. Imported	5. V-8	5. Black

HORSEPOWER	PRICE RANGE	BODY STYLE
1. Below 100	1. Below $1000	1. Sedan
2. 100–199	2. $1000 to $2499	2. Hardtop
3. 200–299	3. $2500 to $3999	3. Convertible
4. 300–399	4. $4000 to $5499	4. Station wagon
5. 400 or higher	5. $5500 or higher	5. Sport van

Also: NEW, 1; USED, 0

The next thing to do is to prepare a DATA statement for each of the cars being considered. These DATA statements will be put at the end of our computer "car-matching" program. We'll show this program shortly.

Here's what these statements might look like:

```
Car #910:  860  DATA 1,910,1,1,5,2,2,2
Car #920:  870  DATA 0,920,4,2,2,3,2,1
```

New or Used | Car # | Make | Engine | Color | HP | Price | Style

Here's a RUN of our program by a customer who does not insist on having a new car. Notice that the customer also rates the importance of each choice according to the following scale:

5 = VERY IMPORTANT TO ME
4 = IMPORTANT TO ME
3 = PREFERRED IF POSSIBLE
2 = WILLING TO CHANGE
1 = NOT REALLY IMPORTANT

```
RUN


HONEST HORACE'S COMPUTERIZED CAR SELECTOR
DO YOU REQUIRE A NEW CAR (1=YES,0=NO)?0

TYPE ONLY A NUMBER FROM 1 TO 5 FOR THE FOLLOWING:
MAKE PREFERRED?1
   IMPORTANCE OF THIS?3
ENGINE PREFERRED?3
   IMPORTANCE OF THIS?3
COLOR PREFERRED?2
   IMPORTANCE OF THIS?3
HORSEPOWER PREFERRED?2
   IMPORTANCE OF THIS?4
PRICE RANGE PREFERRED?3
   IMPORTANCE OF THIS?4
BODY STYLE PREFERRED?1
   IMPORTANCE OF THIS?3

YOUR BEST SELECTION IS CAR # 970.  SELECTION INDEX IS 16.5.
SECOND SELECTION IS CAR #940.  SELECTION INDEX IS 14.
THIRD SELECTION IS CAR # 920.  SELECTION INDEX IS 12.25.

IF YOU DON'T LIKE THESE CHOICES, TRY RUNNING THIS PROGRAM AGAIN
 WITH DIFFERENT 'IMPORTANCE' NUMBERS.
```

The program calculates a *selection index* (c) by a formula we invented for this program. The idea of the formula is to calculate a score for each characteristic and then to add up all these scores. We want a formula which gives a *larger* value for a *closer* match.

Here's how our formula works in calculating the selection index between the customer represented in the RUN above and car #970:

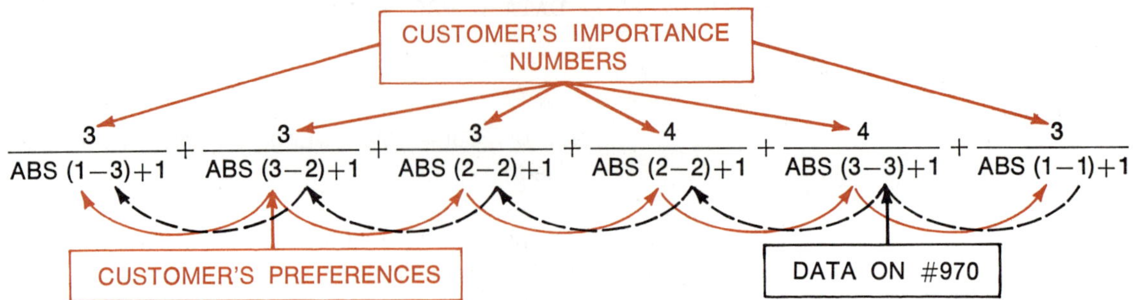

These add up to $\frac{3}{3} + \frac{3}{2} + \frac{3}{1} + \frac{4}{1} + \frac{4}{1} + \frac{3}{1} = 16.5$ as found in the RUN.

You should notice two things about our formula:

1. Because the importance number is in the *numerator* of each term, a person who votes something very important (5) *increases* the value of that characteristic in the match index. In other words, each term is *DIRECTLY* proportional to the importance number.
2. Because the *difference* (in absolute value) between two *characteristics* is in the *denominator,* a match in characteristics (which produces a small difference) causes that term to become large. In other words, each term is *INVERSELY* proportional to: (the difference in characteristics +1). The +1 is included to prevent division by zero. For example:

$$\frac{3}{\text{ABS}(5-1)+1} = 0.6$$ MISMATCH

but

$$\frac{3}{\text{ABS}(5-5)+1} = 3.0$$ MATCH

Here is a listing of program 4E-1. The program will print out the 3 best matches.

```
10  PRINT "HONEST HAL'S COMPUTERIZED CAR SELECTOR"
20  PRINT "DO YOU REQUIRE A NEW CAR (1=YES,0=NO)";
30  INPUT S
40  PRINT
50  PRINT "TYPE ONLY A NUMBER FROM 1 TO 5 FOR";
60  PRINT " THE FOLLOWING:"
70  FOR I=1 TO 6
80  GOTO I OF 90,110,130,150,170,190
90  PRINT "MAKE ";
100  GOTO 200
110  PRINT "ENGINE ";
120  GOTO 200
130  PRINT "COLOR ";
140  GOTO 200
150  PRINT "HORSEPOWER ";
160  GOTO 200
170  PRINT "PRICE RANGE ";
180  GOTO 200
```

Here 0 means that customer will take a new *or* a used car.

Loop for inputting customer's preferences. (Lines 70-340)

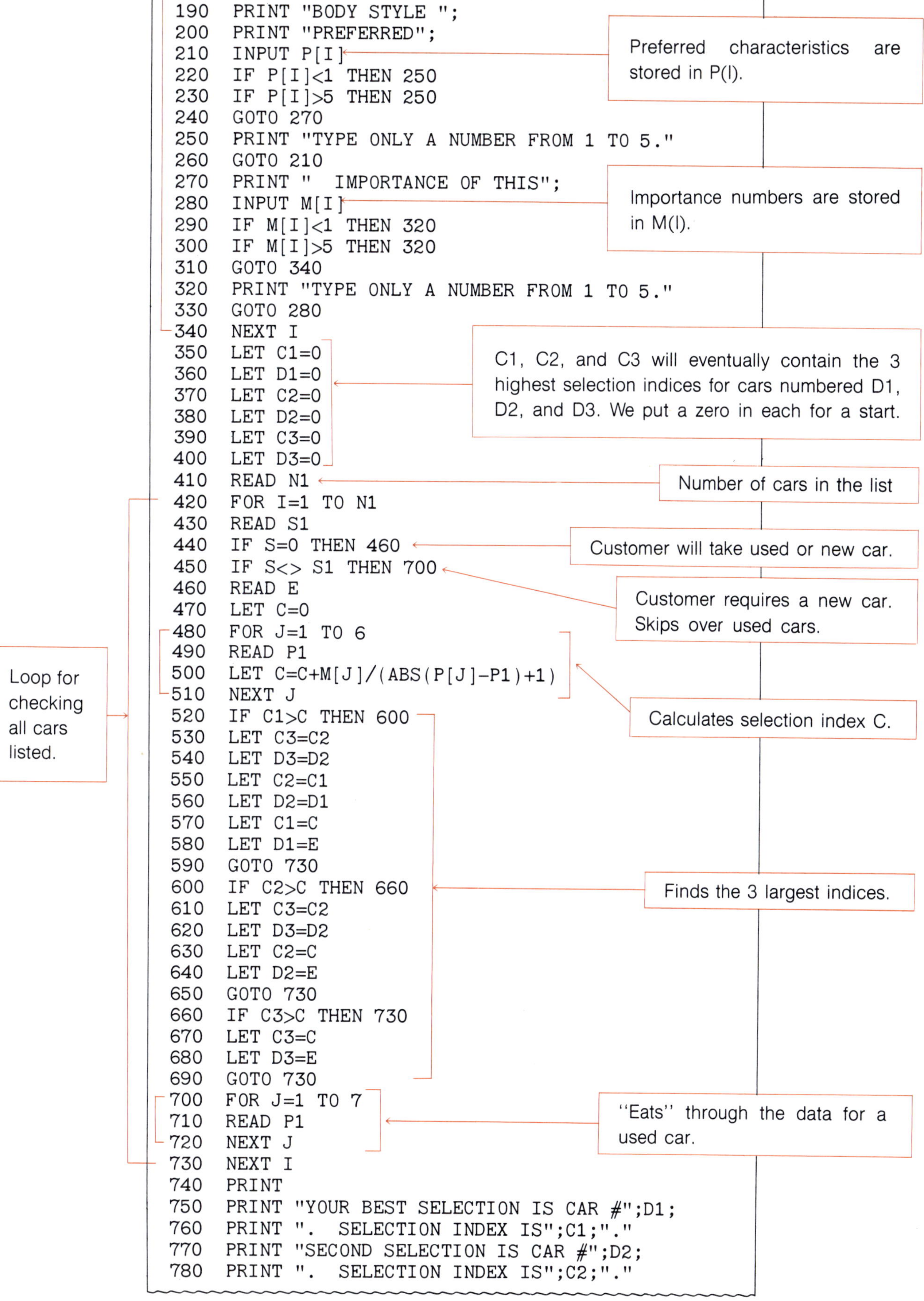

```
190 PRINT "BODY STYLE ";
200 PRINT "PREFERRED";
210 INPUT P[I]
220 IF P[I]<1 THEN 250
230 IF P[I]>5 THEN 250
240 GOTO 270
250 PRINT "TYPE ONLY A NUMBER FROM 1 TO 5."
260 GOTO 210
270 PRINT "  IMPORTANCE OF THIS";
280 INPUT M[I]
290 IF M[I]<1 THEN 320
300 IF M[I]>5 THEN 320
310 GOTO 340
320 PRINT "TYPE ONLY A NUMBER FROM 1 TO 5."
330 GOTO 280
340 NEXT I
350 LET C1=0
360 LET D1=0
370 LET C2=0
380 LET D2=0
390 LET C3=0
400 LET D3=0
410 READ N1
420 FOR I=1 TO N1
430 READ S1
440 IF S=0 THEN 460
450 IF S<> S1 THEN 700
460 READ E
470 LET C=0
480 FOR J=1 TO 6
490 READ P1
500 LET C=C+M[J]/(ABS(P[J]-P1)+1)
510 NEXT J
520 IF C1>C THEN 600
530 LET C3=C2
540 LET D3=D2
550 LET C2=C1
560 LET D2=D1
570 LET C1=C
580 LET D1=E
590 GOTO 730
600 IF C2>C THEN 660
610 LET C3=C2
620 LET D3=D2
630 LET C2=C
640 LET D2=E
650 GOTO 730
660 IF C3>C THEN 730
670 LET C3=C
680 LET D3=E
690 GOTO 730
700 FOR J=1 TO 7
710 READ P1
720 NEXT J
730 NEXT I
740 PRINT
750 PRINT "YOUR BEST SELECTION IS CAR #";D1;
760 PRINT ".  SELECTION INDEX IS";C1;"."
770 PRINT "SECOND SELECTION IS CAR #";D2;
780 PRINT ".  SELECTION INDEX IS";C2;"."
```

```
790  PRINT "THIRD SELECTION IS CAR #";D3;
800  PRINT ".  SELECTION INDEX IS";C3;"."
810  PRINT
820  PRINT "IF YOU DON'T LIKE THESE CHOICES,";
830  PRINT " TRY RUNNING THIS PROGRAM AGAIN";
840  PRINT " WITH DIFFERENT 'IMPORTANCE' NUMBERS."
850  DATA 9
860  DATA 1,910,1,1,5,2,2,2
870  DATA 0,920,4,2,2,3,2,1
880  DATA 1,930,2,4,5,1,2,2
890  DATA 0,940,3,5,2,2,2,1
900  DATA 1,950,3,5,3,4,4,4
910  DATA 1,960,5,3,5,4,5,2
920  DATA 1,970,3,2,2,2,3,1
930  DATA 1,980,2,2,3,2,2,2
940  DATA 0,990,3,2,1,3,3,5
950  END
```

4E-2: WEIGH-MATCH Experiment with other formulas for the match. For example, squaring the numerator in each term of the formula will give more "weight" to the importance number.

4E-3: FILES If your computer has file commands, and you know how to use them, put the data on files instead of in DATA commands. (You're on your own for this problem. Your computer reference manual should be consulted.)

SECTION 5 OPEN SENTENCES IN TWO VARIABLES

Checklist of Computing Skills

Previously explained:	PRINT, END, LET, INPUT, IF . . . THEN, GOTO, FOR, NEXT, STEP, INT, PRINT TAB, subscripted variables—single subscripts
Explained in this section:	GOSUB & RETURN (Explained in Unit 5B) Sorting Routine (Used in Unit 5B)

Checklist of Algebraic Skills

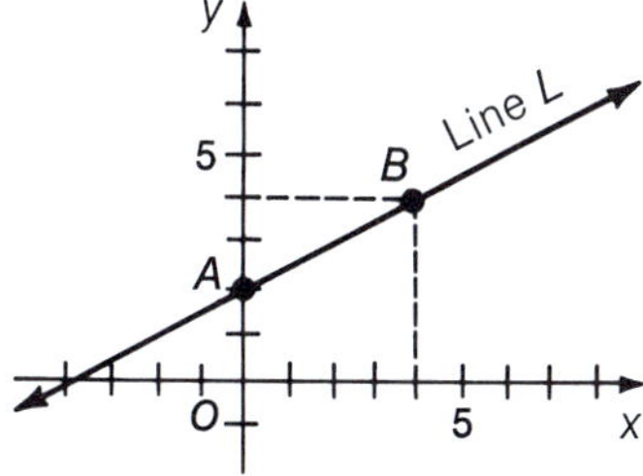

1. The coordinates of point A are _?_, _?_.
2. The coordinates of point B are _?_, _?_.
3. The slope of line L is _?_.
4. The y-intercept of L is _?_.
5. The equation of line L is $y =$ _?_ $\times x +$ _?_.

(Needed in Unit 5E) What are the missing values? The straight line through (0, 1) and (2, 5) has the equation: $y =$ _?_ $\times x +$ _?_ (6) (7)

The straight line through (−1, 5) and (0, 0) has the equation:

$y =$ _?_ $\times x +$ _?_ (8) (9)

Check your answers with those at the bottom of the page.

Unit 5A. The Automated Dart Thrower

Below you see pictured a square dart board superimposed on a coordinate system. In this unit we'll program the computer to "throw darts" at a board

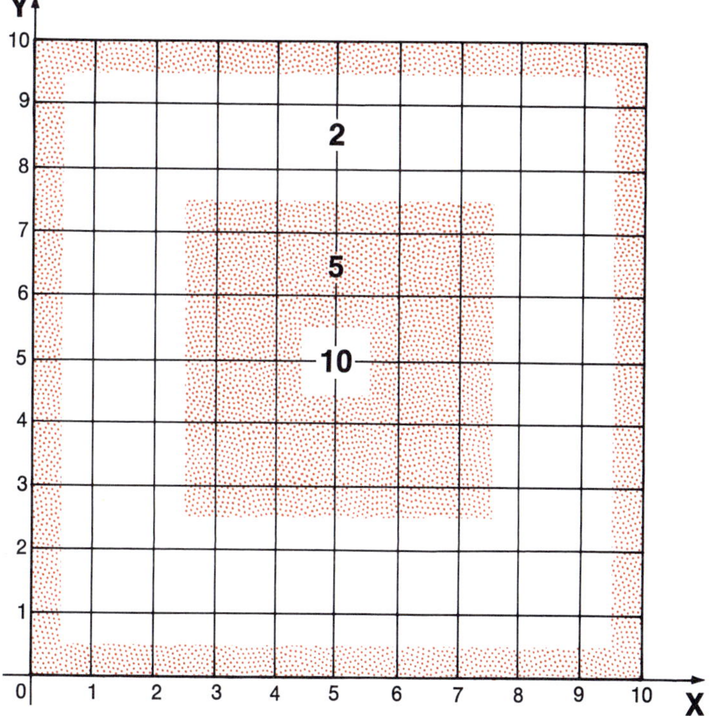

Answers: (1) 0, 2 (2) 4, 4 (3) 0.5 (4) 2 (5) $y = 0.5 \times x + 2$ (6) 4 (7) 1 (8) −5 (9) 0

and tell us where they landed. To make things simple, the computer will use only integers to describe the position of a dart.

The position of a dart will be printed out in the form (3, 4). This means that the dart landed on the board at the point where the X = 3 line intersects the Y = 4 line.

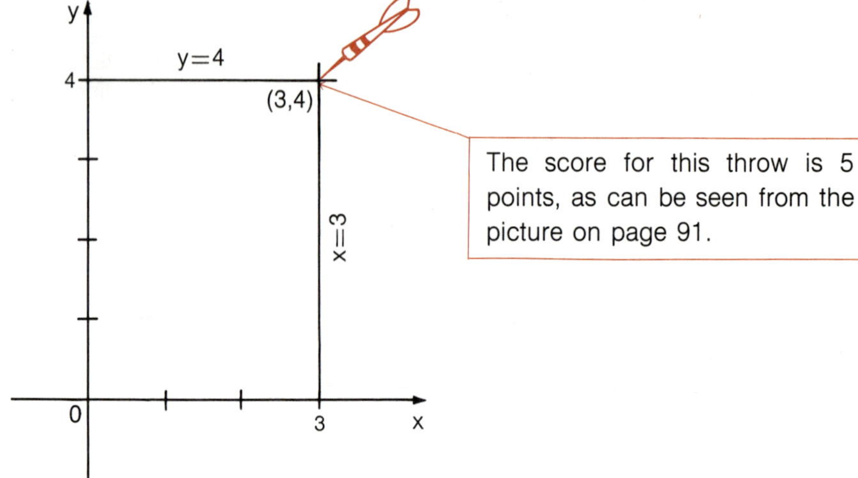

5A-1: DART-THROW This program "throws darts" by using randomly selected integers and prints a final score as explained above.

The object of the game is to get as high a score as possible. You can try to influence the dart throwing by typing in a lucky number. Your lucky number is used to randomize the random generator as explained in experiment [3] of unit 4A.

Here's a sample RUN of such a program:

```
RUN


TYPE THE NUMBER OF DARTS YOU WISH (BETWEEN 2 AND 10):
?5
TYPE IN YOUR LUCKY 2-DIGIT NUMBER BETWEEN 10 AND 99:
?35
DART  1  AT  ( 4, 7)
DART  2  AT  ( 5, 7)
DART  3  AT  ( 7, 6)
DART  4  AT  ( 5, 9)
DART  5  AT  ( 9, 2)


YOUR TOTAL POINT SCORE FOR THE 5 DARTS IS 19.
```

A listing of the program that gave the preceding RUN is shown on the next page.

```
5  PRINT "TYPE THE NUMBER OF DARTS YOU WISH (BETWEEN 2 AND 10):"
10  INPUT N
15  IF N<2 THEN 5
20  IF N>10 THEN 5
25  PRINT "TYPE IN YOUR LUCKY 2-DIGIT NUMBER BETWEEN 10 AND 99;"
30  INPUT L
35  IF L<10 THEN 25
40  IF L>99 THEN 25
45  LET Q=RND(-1)
50  FOR I=1 TO L
55  LET Q=RND(1)
60  NEXT I
65  FOR I=1 TO N
70  LET X[I]=INT(9*RND(1)+1)
75  LET Y[I]=INT(9*RND(1)+1)
80  PRINT "DART ";I;" AT ("X[I];",";Y[I];")"
85  NEXT I
90  PRINT
95  PRINT
100  LET S=0
105  FOR I=1 TO N
110  IF X[I]<2·5 THEN 160
115  IF Y[I]<2·5 THEN 160
120  IF X[I]>7·5 THEN 160
125  IF Y[I]>7·5 THEN 160
130  IF X[I]<4·5 THEN 155
135  IF Y[I]<4·5 THEN 155
140  IF X[I]>5·5 THEN 155
145  IF Y[I]>5·5 THEN 155
150  LET S=S+5
155  LET S=S+3
160  LET S=S+2
165  NEXT I
170  PRINT "YOUR TOTAL POINT SCORE FOR THE";N;"DARTS IS";S;"."
175  END
```

Use this line or omit it as required by your system [line 45]

5A-2: DART-SCORE Improve program 5A-1 so that each dart has its score printed, for example:

DART 1 AT (4,8) (2 POINTS)

5A-3: DART-GAME Write a computer-vs.-person dart game with the following rules:

(1) Players should take turns, with the computer first throwing a dart randomly, followed by the person throwing his dart.

(2) The *sum* of the person's coordinates must be the same as the sum of the computer's for the person to score. Otherwise, the computer scores.

(3) Each dart must be more than one unit distance from any previous dart. If not, provide a penalty.

Unit 5B. The Automated Dart Board

5B-1: DART-BOARD Add a dart-board picture to program 5A-1 so that a RUN looks like this:

```
RUN


TYPE THE NUMBER OF DARTS YOU WISH (BETWEEN 2 AND 10):
?5
TYPE IN YOUR LUCKY 2-DIGIT NUMBER BETWEEN 10 AND 99:
?35
DART  1  AT  ( 4, 7)
DART  2  AT  ( 5, 7)
DART  3  AT  ( 7, 6)
DART  4  AT  ( 5, 9)
DART  5  AT  ( 9, 2)

 10 +
    :
  9 +                  X
    :
  8 +
    :         . . . . . . . . . . .
  7 +                X   X
    :         :                   :
  6 +                        X
    :         :       . . .       :
  5 +
    :         :       . . .       :
  4 +
    :         :                   :
  3 +
    :         . . . . . . . . . . .
  2 +                                    X
    :
  1 +
    :
  0 +---+---+---+---+---+---+---+---+---+---+
    0   1   2   3   4   5   6   7   8   9   10

YOUR TOTAL POINT SCORE FOR THE 5 DARTS IS 19.
```

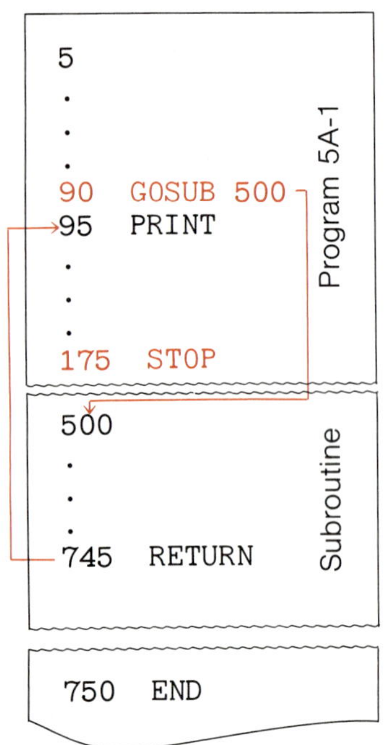

The way we suggest doing this is to write the "picture" program as a subroutine, using the GOSUB and RETURN statements.

NEW: GOSUB, RETURN

The GOSUB statement sends the execution of the program to the subroutine. The RETURN statement ends the subroutine and takes the program back to the line right after GOSUB.

To make program 5B-1, you add the subroutine program beginning with line 500 and change lines 90 and 175 in program 5A-1 to:

```
90 GOSUB 500
175 STOP
```

Here is a listing of a subroutine for "drawing" the dart board.

```
500  FOR I1=1 TO N-1
505  LET C=N+1-I1
510  FOR I=2 TO C
515  IF Y[I-1] >= Y[I] THEN 550
520  LET Y1=Y[I-1]
525  LET X1=X[I-1]
530  LET Y[I-1]=Y[I]
535  LET X[I-1]=X[I]
540  LET Y[I]=Y1
545  LET X[I]=X1
550  NEXT I
555  NEXT I1
560  FOR I1=1 TO N-1
565  LET C=N+1-I1
570  FOR I=2 TO C
575  IF Y[I-1] <> Y[I] THEN 600
580  IF X[I-1] <= X[I] THEN 600
585  LET X1=X[I-1]
590  LET X[I-1]=X[I]
595  LET X[I]=X1
600  NEXT I
605  NEXT I1
610  PRINT
615  LET I=1
620  FOR J=10 TO 1 STEP -1
625  PRINT J;TAB(4);"+";
630  IF I>N THEN 655
635  IF J<> Y[I] THEN 655
640  PRINT TAB(4+4*X[I]);"X";
645  LET I=I+1
650  GOTO 630
655  PRINT
660  PRINT TAB(4);":";
665  IF J=8 THEN 705
670  IF J=3 THEN 705
675  IF J=6 THEN 715
680  IF J=5 THEN 715
685  IF J>8 THEN 725
690  IF J<3 THEN 725
695  PRINT TAB(6+4*2);":";TAB(6+4*7);":"
700  GOTO 730
705  PRINT TAB(6+4*2);". . . . . . . . . . ."
710  GOTO 730
715  PRINT TAB(6+4*2);":";TAB(6+4*4);". . .";TAB(6+4*7);":"
720  GOTO 730
725  PRINT
730  NEXT J
735  PRINT 0;TAB(4);"+---+---+---+---+---+---+---+---+---+---+"
740  PRINT TAB(4);"0   1   2   3   4   5   6   7   8   9   10"
745  RETURN
```

These loops *sort* the pairs of coordinates in descending order of the Y-coordinates. (lines 500–555)

If several pairs have the same Y-coordinate, these loops *sort* the pairs of coordinates in ascending order of the X-coordinates. (lines 560–605)

The best way to see how the sorting routines work is to run through them "by hand," using the data of our sample RUN.

W

25 cm 25 cm

W

Unit 5C. Graphs by Computer

Although it has its limitations, a computer terminal can be used to graph ordered pairs of numbers in a coordinate plane. The dart board shown in Unit 5B is an example of what the results can look like.

In this unit we are going to see how we can graph the set of ordered pairs (a function) described by some *formula*. For example, the perimeter of a rectangle 25 centimeters high can be expressed in terms of its width as:

$$P = 2W + 50$$

Corresponding values of W and P can be listed in a table:

W	P
1	52
2	54
3	56
⋮	⋮

Producing such a table is an easy task for a computer:

```
10  PRINT " W"," P"
20  FOR W=1 TO 10
30  PRINT W,2*W+50
40  NEXT W
50  END
```

It would be even better to have a picture—a graph—of this function. Here's how we can do it.

Step 1. Choose a domain for the function: $\{A \leq W \leq B\}$ $(A \neq B)$
Step 2. Choose a step size D.
Step 3. Assign the value of A to the variable W.
Step 4. Calculate P with the given formula.
Step 5. Place a "*" at the point (W, P) on the graph.
Step 6. Increase W by the step size D.
Step 7. If $W \leq B$ go to step 4; otherwise stop.

Note that steps 4 through 7 are repeated; hence P is computed and the corresponding points graphed for $W = A$, $A + D$, $A + 2{*}D$, . . . , until $W > B$.

We can write a computer program which follows these steps. A set of steps like this is called an *algorithm;* an algorithm is a recipe for doing something.

5C-1: GRAPH Here is a program for graphing a portion of the function defined by:

$$P = 2W + 50$$

The values of W will run vertically, and the values of P will run horizontally. If we start W with the value 1, the corresponding value of P will be 52.

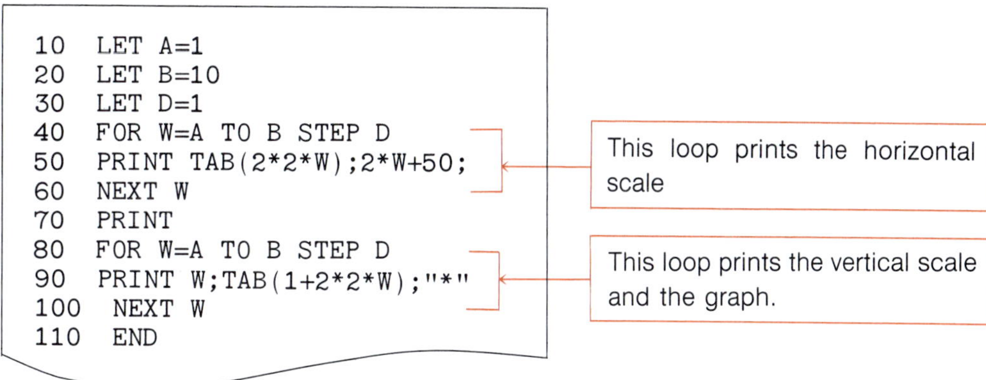

```
10  LET A=1
20  LET B=10
30  LET D=1
40  FOR W=A TO B STEP D
50  PRINT TAB(2*2*W);2*W+50;
60  NEXT W
70  PRINT
80  FOR W=A TO B STEP D
90  PRINT W;TAB(1+2*W);"*"
100  NEXT W
110  END
```


Look at the TAB values in lines 50 and 90. The expression 2 * W is multiplied by 2 so that 2 horizontal spaces will represent a unit. This will make the grid nearly square when a line feed represents 1 vertical unit. In line 90, 1 is added to the TAB value so that each * will align with the corresponding value on the scale.

Below is a RUN of this program.

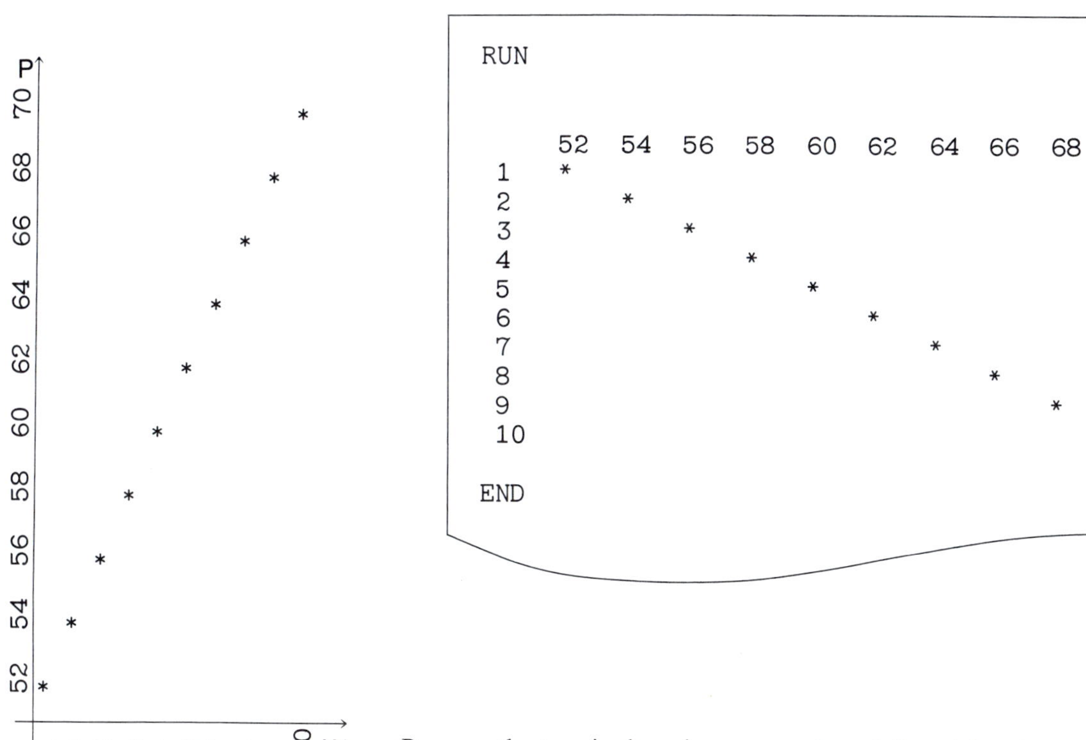

```
RUN

    52  54  56  58  60  62  64  66  68  70
 1  *
 2      *
 3          *
 4              *
 5                  *
 6                      *
 7                          *
 8                              *
 9                                  *
 10                                     *

END
```


Because the terminal carriage moves from left to right when graphing P (the "Y" axis), the graph in the RUN above is sideways. The picture above left shows how it should be rotated to look like the graphs in your algebra book. Notice that the W and P axes have been added to the graph.

Unit 5D. Straight-Line Graphs

As you know by now, the graphs "drawn" on a terminal are only rough approximations. However, such graphs are very useful, so let's take a look at how we can make them as accurate as possible.

We first consider the equation of a straight line in the form

$$Y = MX + B$$

where M is the slope and B is the Y-intercept. Later on we will write programs which plot the graphs of nonlinear functions.

5D-1: LINE-GRAPH1 Write a program that graphs an equation of a straight line

$$Y = M * X + B$$

when M and B are INPUT. (Assume that $M \neq 0$.) The program should also allow users to type in the maximum and minimum values they wish for X. (Assume that these values are distinct.) Pattern the scales after those used in printing the dart board in program 5B-1, but arrange them so that the horizontal axis (Y-axis) is at the top (as in program 5C-1) and the vertical scale increases from top to bottom.

The sample solution on the next page shows one way of writing such a program. You can draw a horizontal line through 0 on the vertical scale and a vertical line through the value of the Y-intercept on the horizontal scale to verify that the Y-intercept is the one specified.

```
10  PRINT "WHAT IS THE SLOPE OF YOUR LINE (NOT 0)";
15  INPUT M
20  PRINT "AND THE Y-INTERCEPT";
25  INPUT B
30  PRINT "WHAT IS THE MINIMUM VALUE X WILL HAVE";
35  INPUT X1
40  PRINT "AND THE MAXIMUM";
45  INPUT X2
50  PRINT
55  LET Y1=M*X1+B
60  LET Y2=M*X2+B
65  IF M>0 THEN 85
70  LET Y3=Y1
75  LET Y1=Y2
80  LET Y2=Y3
85  IF Y2-Y1>15 THEN 185
90  FOR I=Y1 TO Y2
95  PRINT TAB(5+4*(I-Y1));I;
100  NEXT I
105  PRINT
110  PRINT TAB(3);"---+";
115  FOR I=1 TO (Y2-Y1)
120  PRINT "---+";
125  NEXT I
130  PRINT " Y"
135  FOR X=X1 TO X2 STEP .5
140  LET Y=M*X+B
145  IF X <> INT(X) THEN 160
150  PRINT X;TAB(3);"+";
155  GOTO 165
160  PRINT TAB(3);":";
165  PRINT TAB(6+4*(Y-Y1));"*"
170  NEXT X
175  PRINT "  X"
180  STOP
185  PRINT "RANGE TOO LARGE"
190  END
RUN
```

These loops print the horizontal scale. (lines 90–130)

This loop prints the vertical scale and the graph. (lines 135–175)

```
WHAT IS THE SLOPE OF YOUR LINE (NOT 0)?2
AND THE Y-INTERCEPT?1
WHAT IS THE MINIMUM VALUE X WILL HAVE?-2
AND THE MAXIMUM?2

     -3  -2  -1   0   1   2   3   4   5
   ---+---+---+---+---+---+---+---+---+ Y
-2 +  *
   :      *
-1 +          *
   :              *
 0 +                  *
   :                      *
 1 +                          *
   :                              *
 2 +                                  *
   X
```

The preceding program gives graphs of linear equations provided that the range is not too great. On the other hand, it is possible to program the computer to adjust the scales so that graphs covering any range can be plotted on a computer terminal.

5D-2: LINE-GRAPH2 Write a program that will compute the scales so that the graph always fills 61 spaces in the Y direction and 41 lines in the X direction. Here's a sample solution:

```
10  PRINT "WHAT IS THE SLOPE OF YOUR LINE (NOT 0)";
15  INPUT M
20  PRINT "AND THE Y-INTERCEPT";
25  INPUT B
30  PRINT "WHAT IS THE MINIMUM VALUE X WILL HAVE";
35  INPUT D1
40  PRINT "AND THE MAXIMUM";
45  INPUT D2
50  PRINT
55  LET R1=M*D1+B
60  LET R2=M*D2+B
65  LET K1=(41-1)/(D2-D1)
70  LET K2=(61-1)/(R2-R1)
75  LET J1=21-K1*(D2+D1)/2
80  LET J2=31-K2*(R2+R1)/2
85  FOR I=1 TO 61 STEP 10
90  PRINT TAB(5+I);INT(10*(I-J2)/K2+.5)/10;
95  NEXT I
100  PRINT
105  FOR I=1 TO 61 STEP 10
110  PRINT TAB(6+I);"↑";
115  NEXT I
120  PRINT
125  PRINT TAB(5);"--";
130  FOR I=1 TO 61
135  PRINT "-";
140  NEXT I
145  PRINT " Y"
150  LET X=D1
155  FOR I=1 TO 41
160  LET Y=M*X+B
165  IF (I-1)/10 <> INT((I-1)/10) THEN 180
170  PRINT INT(10*(I-J1)/K1+.5)/10;TAB(4);"+";
175  GOTO 185
180  PRINT TAB(4);":";
185  PRINT TAB(6+K2*Y+J2+.5);"*"
190  LET X=X+1/K1
195  NEXT I
200  PRINT "    X"
205  END
```


Notice that this program places scale numbers on every 10th line and every 10th space.

The .5 is added in lines 90, 170, and 185 to give rounded values.

Here's a sample RUN.

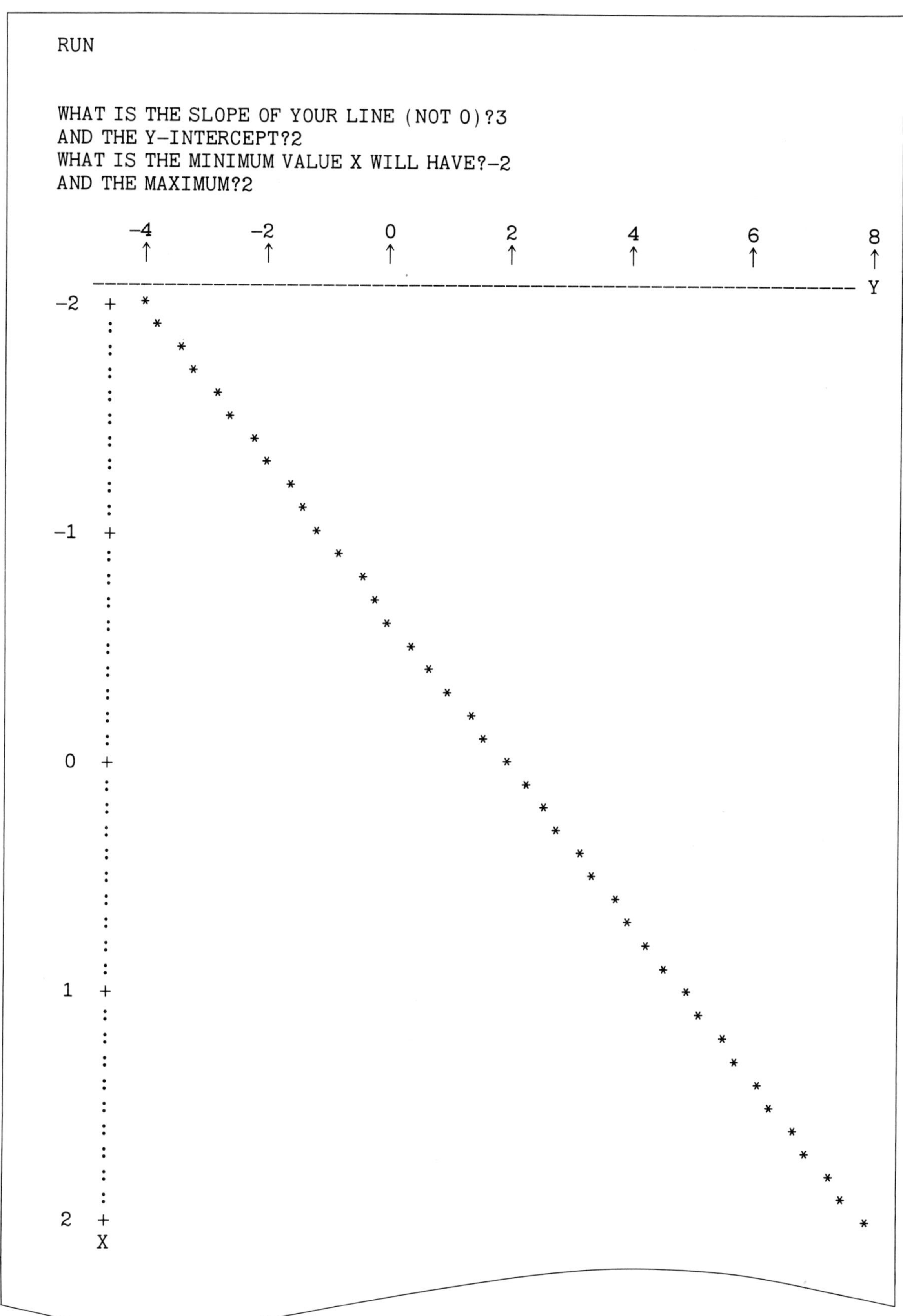

A partial RUN with different INPUT numbers is shown on page 102.

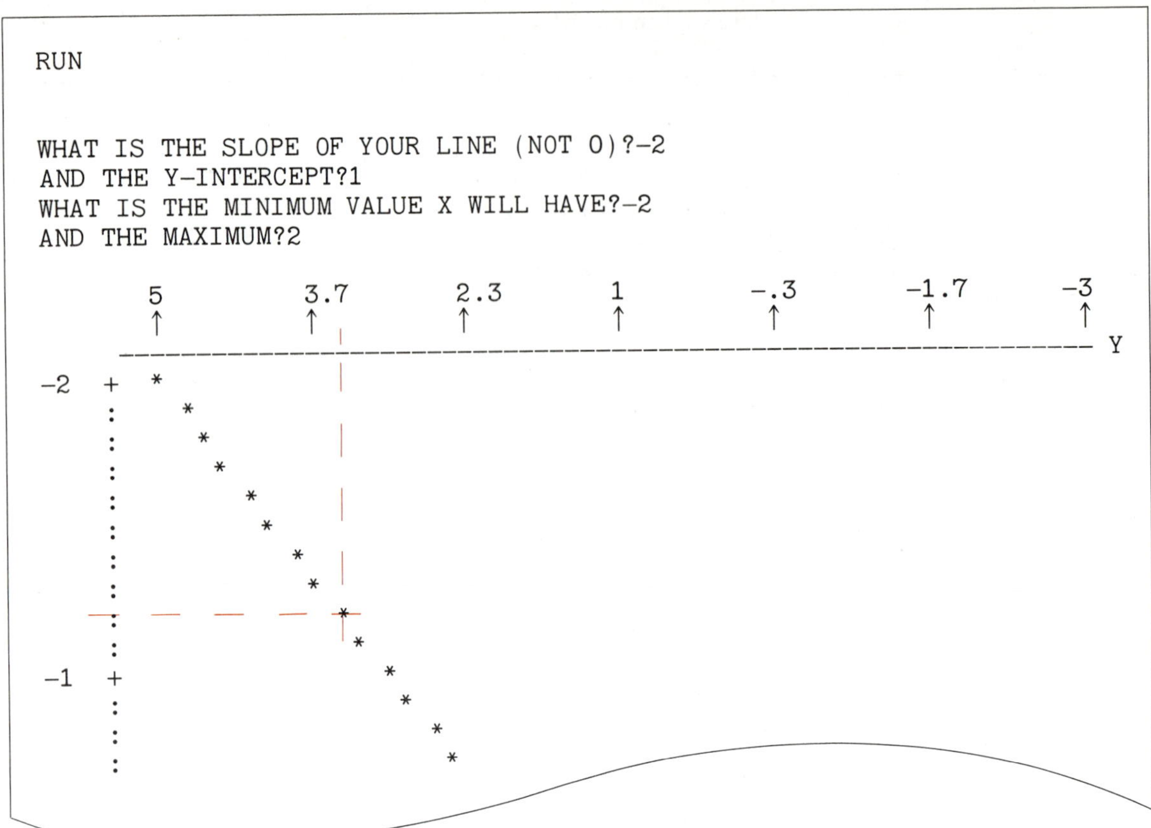

```
RUN


WHAT IS THE SLOPE OF YOUR LINE (NOT 0)?-2
AND THE Y-INTERCEPT?1
WHAT IS THE MINIMUM VALUE X WILL HAVE?-2
AND THE MAXIMUM?2
         5         3.7        2.3         1         -.3        -1.7        -3
         ↑          ↑          ↑          ↑          ↑          ↑          ↑
      --------------------------------------------------------------------------- Y
-2   +  *
     :    *
     :     *
     :      *
     :        *
     :         *
     :           *
     :            *
     :               *
     :                *
     :
-1   +                  *
     :                   *
     :                     *
     :                      *
```

Notice that the graphs of all lines that you obtain with this program will look practically the same! Only the scales will be changed. When the slope is negative, the direction on the Y-axis is reversed.

To read off corresponding values of X and Y, you must compute the intermediate value on each scale. [This is an example of *linear interpolation* (see Unit 5E).] For example, the coordinates of the * at the intersection of the dashed lines are approximately $(-2 + .8(-1 - (-2)), 2.3 + .8(3.7 - 2.3))$ or $(-1.2, 2.42)$.

5D-3: LINE-GRAPH3 In earlier programs, we have drawn the X and Y scales off to the side. Can you make the program draw the axes through the origin (0, 0)? If you need help, look ahead at the programs in Section 8D. Also see if you can prevent reversal of the Y-axis for negative slopes.

Unit 5E. Linear Fortune Tellers

The graph on page 103 shows how many liters (L) of antifreeze to add to a car radiator for protection against various lowest possible temperatures.

The dotted lines show, for example, that 1 liter is needed if the lowest temperature expected is 15 degrees below zero Celsius (centigrade), that is, $T = -15$. We shall assume that no antifreeze is needed when the temperature does not fall below zero. Therefore, two points on the graph are:

$$(-15, 1) \text{ and } (0, 0)$$

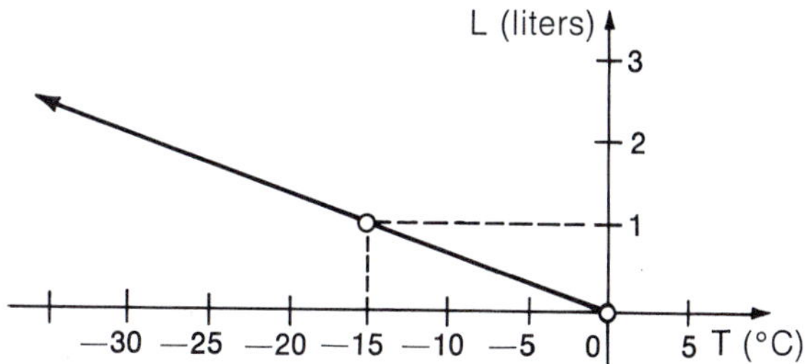

If we assume that the graph is a straight line, that is, that L = MT + B, then we can predict the number of liters needed for any other temperature T below zero. This is our kind of "fortune telling." If T is between the given values, this is called **linear interpolation.** If T is not between the given values, it is called **linear extrapolation.**

5E-1: ANTIFREEZE Write a program (based on the above graph) which will draw a graph of the function defined by L = MT + B for the domain $\{T: -40 \leq T \leq 0\}$. Your program must first figure out the values of M and B from the given values of (T1,L1) and (T2,L2).

5E-2: INTERP Write a program that does the following:

```
RUN


INPUT A KNOWN PAIR OF COORDINATES: T1,L1?-15,1
ANOTHER PAIR: T2,L2?0,0
INPUT A TEMPERATURE:?-10
BY LINEAR INTERPOLATION, YOU'LL NEED .666667 LITERS.

END
RUN


INPUT A KNOWN PAIR OF COORDINATES: T1,L1?-15,1
ANOTHER PAIR: T2,L2?0,0
INPUT A TEMPERATURE:?-20
BY LINEAR EXTRAPOLATION, YOU'LL NEED 1.33333 LITERS.

END
```

Be sure to include an appropriate response if a positive value of T is input.

> NOTE: 1 liter = 2.11344 pints
> Values of temperature in Fahrenheit degrees can be found from the values in Celsius (centigrade) degrees by using: $F = \frac{9}{5}C + 32$

5E-3: LINEAR-INTERP Generalize program 5E-2 to allow input of any values of (x_1,y_1) and (x_2,y_2). Look ahead to Unit 6D if you need ideas.

SECTION 6 SYSTEMS OF LINEAR EQUATIONS

Checklist of Computing Skills

Previously explained:	PRINT, END, LET, INPUT, IF . . . THEN, STOP, GOTO, FOR, NEXT, STEP, SQR, ABS, INT, PRINT TAB, subscripted variables—single subscripts, GOSUB, RETURN
Explained in this section:	Subscripted variables—double subscripts

Checklist of Algebraic Skills

A. Solve this system: $2x + y = 11$ $\quad x =$ ___?___ (1)
$\quad\quad x - y = 1$ $\quad y =$ ___?___ (2)

B. The system $\quad 5 * X + 15 * Y = 10$
$\quad\quad 3 * Y + 2 * Y = -8$

is equivalent to the system: $1 * X +$ ___?___ $* Y =$ ___?___ (3) (4)
$\quad\quad 0 * X +$ ___?___ $* Y =$ ___?___ (5) (6)

(The following questions relate only to units 5D and 5E.)

C. For the line through the points (2, 3) and (−1, 0), the slope is ___?___, and the *y*-intercept is ___?___. (7) (8)

D. For the line through the points (4, −1) and (2, 1), the slope is ___?___, and the *y*-intercept is ___?___. (9) (10)

E. The above two lines intersect at (___?___, ___?___). (11) (12)

Check your answers with those at the bottom of the page.

NEW: BASIC VARIABLES WITH 2 SUBSCRIPTS

Suppose that you are in charge of sales for a newspaper, and you want to know on which *street* (or which *avenue*) the most papers are sold, but the only information you have is on how many papers are sold at corner newstands. The array below shows how many papers are sold at each *corner*.

	1st Avenue	2d Avenue	3d Avenue	4th Avenue
1st Street	103	48	0	289
2d Street	88	51	310	0
3d Street	0	72	98	243

For example, this number means that 72 papers are sold at the corner of 3d Street and 2d Avenue.

In BASIC, we are allowed to store this number (72) in a variable called N(3, 2). This is a *variable with two subscripts.*

Answers: (1) 4 (2) 3 (3) 3 (4) 2 (5) −7 (6) −14 (7) 1 (8) 1 (9) −1 (10) 3 (11) 1 (12) 2

The variable N(S, A) is the name of the computer location in which we have decided to store the *number* of papers sold at street S and avenue A. There will be *twelve* of these variables, one for each street corner:

N(1, 1)	N(1, 2)	N(1, 3)	N(1, 4)
N(2, 1)	N(2, 2)	N(2, 3)	N(2, 4)
N(3, 1)	N(3, 2)	N(3, 3)	N(3, 4)

You can use nested loops to help you input these values into your program. For example:

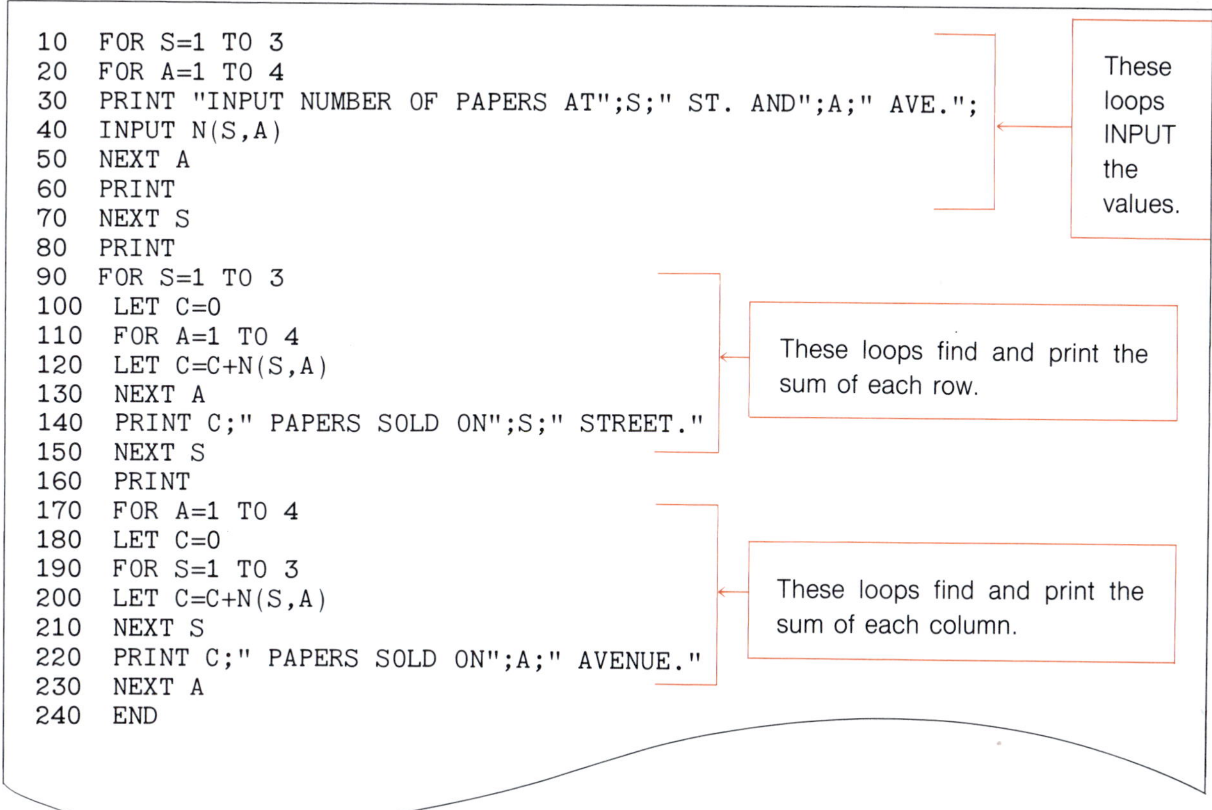

```
10  FOR S=1 TO 3
20  FOR A=1 TO 4
30  PRINT "INPUT NUMBER OF PAPERS AT";S;" ST. AND";A;" AVE.";
40  INPUT N(S,A)
50  NEXT A
60  PRINT
70  NEXT S
80  PRINT
90  FOR S=1 TO 3
100  LET C=0
110  FOR A=1 TO 4
120  LET C=C+N(S,A)
130  NEXT A
140  PRINT C;" PAPERS SOLD ON";S;" STREET."
150  NEXT S
160  PRINT
170  FOR A=1 TO 4
180  LET C=0
190  FOR S=1 TO 3
200  LET C=C+N(S,A)
210  NEXT S
220  PRINT C;" PAPERS SOLD ON";A;" AVENUE."
230  NEXT A
240  END
```


In lines 90 to 150, we add the number of papers sold along each street (row of our table). The adding is done in the inner loop (lines 110–130):

FIRST: We set S = 1 (line 90) and add:
N(1, 1) + N(1, 2) + N(1, 3) + N(1, 4)
THEN: We set S = 2 (line 90) and add:
N(2, 1) + N(2, 2) + N(2, 3) + N(2, 4)
FINALLY: We set S = 3 (line 90) and add:
N(3, 1) + N(3, 2) + N(3, 3) + N(3, 4)

In lines 170 to 210, we switch the FOR loops around so that we now add the number of papers sold along each avenue (column).

FIRST: N(1, 1) + N(2, 1) + N(3, 1)
THEN: N(1, 2) + N(2, 2) + N(3, 2)
THEN: N(1, 3) + N(2, 3) + N(3, 3)
FINALLY: N(1, 4) + N(2, 4) + N(3, 4)

Now look at the sample RUN on page 106.

```
RUN


INPUT NUMBER OF PAPERS AT 1 ST. AND 1 AVE.?103
INPUT NUMBER OF PAPERS AT 1 ST. AND 2 AVE.?48
INPUT NUMBER OF PAPERS AT 1 ST. AND 3 AVE.?0
INPUT NUMBER OF PAPERS AT 1 ST. AND 4 AVE.?289

INPUT NUMBER OF PAPERS AT 2 ST. AND 1 AVE.?88
INPUT NUMBER OF PAPERS AT 2 ST. AND 2 AVE.?51
INPUT NUMBER OF PAPERS AT 2 ST. AND 3 AVE.?310
INPUT NUMBER OF PAPERS AT 2 ST. AND 4 AVE.?0

INPUT NUMBER OF PAPERS AT 3 ST. AND 1 AVE.?0
INPUT NUMBER OF PAPERS AT 3 ST. AND 2 AVE.?72
INPUT NUMBER OF PAPERS AT 3 ST. AND 3 AVE.?98
INPUT NUMBER OF PAPERS AT 3 ST. AND 4 AVE.?243


 440 PAPERS SOLD ON 1 STREET.
 449 PAPERS SOLD ON 2 STREET.
 413 PAPERS SOLD ON 3 STREET.

 191 PAPERS SOLD ON 1 AVENUE.
 171 PAPERS SOLD ON 2 AVENUE.
 408 PAPERS SOLD ON 3 AVENUE.
 532 PAPERS SOLD ON 4 AVENUE.
```

NOTE: We'll see another use of variables with double subscripts in program 6C-2.

Unit 6A. Simple Linear Systems

6A-1: SIMPLE-SYS Write a program that lets a student practice solving 2 linear equations in 2 variables. Generate the coefficients randomly, but make them simple integers. Give 10 points for correctly finding the value of the first variable, and 5 points for finding the correct value of the second.

Here's what a RUN might look like:

```
RUN


SIMPLE LINEAR SYSTEMS IN 2 VARIABLES:
HOW MANY PROBLEMS?2

PROBLEM 1.  FIND X AND Y WHERE:
    X + 3Y = 0
( 0)X + 2Y =-1
Y =?-.5
VERY GOOD (10 POINTS)
X =?1.5
CORRECT (5 POINTS)
```

$$
\begin{aligned}
3x + 9y &= 3 \\
-3x + y &= 1 \\
\hline
10y &= 4 \\
y &= 0.4 \\
x &= 1 - 3(0.4) \\
&= -0.2
\end{aligned}
$$

```
PROBLEM 2.  FIND X AND Y WHERE:
    X + 3Y = 1
(-3)X + 1Y = 1
Y =?.4
VERY GOOD (10 POINTS)
X =?-.2
CORRECT (5 POINTS)

YOUR TOTAL SCORE IS 30 OUT OF 30 = 100%
```

In our program, the coefficient of X in the first equation is always 1. This makes it easy to solve the system by multiplying both members of the first equation by the negative of the coefficient of X in the second equation, then adding the resulting equations. See the solution to Problem 2, above left.

Here's a listing of our practice program. Line 100 is used to make sure that the randomly generated system will have a unique solution. Notice also that the coefficients of Y are always positive in this program.

```
5 RANDOMIZE (Use only if needed.)
10 PRINT "SIMPLE LINEAR SYSTEMS IN 2 VARIABLES:"
20 PRINT "HOW MANY PROBLEMS";
30 INPUT N
40 PRINT
50 LET P=0
60 FOR I=1 TO N
70 LET Y1=INT(3*RND(1)+1)
80 LET Y2=INT(3*RND(1)+1)
90 LET X2=INT(7*RND(1)-3)
100 IF Y2-X2*Y1=0 THEN 70
110 LET C1=INT(7*RND(1)-3)
120 LET C2=INT(7*RND(1)-3)
130 PRINT "PROBLEM";I;". FIND X AND Y WHERE:"
140 PRINT "    X +";Y1;"Y =";C1
150 PRINT "(";X2;")X +";Y2;"Y =";C2
160 LET Y=(C2-X2*C1)/(Y2-X2*Y1)
170 LET X=C1-Y1*Y
180 PRINT "Y =";
190 INPUT R
200 IF ABS(R-Y)<.01 THEN 230
210 PRINT "NO.  Y =";Y
220 GOTO 250
230 PRINT "VERY GOOD (10 POINTS)"
240 LET P=P+10
250 PRINT "X =";
260 INPUT R
270 IF ABS(R-X)<.01 THEN 300
280 PRINT "NO.  X =";X
290 GOTO 320
300 PRINT "CORRECT (5 POINTS)"
310 LET P=P+5
320 PRINT
330 NEXT I
340 PRINT "YOUR TOTAL SCORE IS";P;" OUT OF";15*N;" =";
350 PRINT P/(15*N)*100;"%"
360 END
```

Exercise: Verify these formulas.

Unit 6B. Tough Linear Systems

Program 6A-1 gave practice in solving linear systems with simple coefficients. But life's not like that—the coefficients of systems arising from real problems are usually not simple integers.

6B-1: TOUGH-SYS Write a program that lets you practice solving linear systems with coefficients that are not integers. Generate the coefficients with the formula:

(INT(199 * RND(1) − 99))/10

Also include a calculator feature in your program (see program 3B-1) so that you can do the required arithmetic right in the computer. Note how the calculator is used in the RUN below.

Here's what a sample RUN might look like:

```
RUN


TOUGH LINEAR SYSTEMS IN 2 VARIABLES:
ONLY COEFFICIENTS WILL BE PRINTED.  FOR EXAMPLE:
2.3   -5.3   4.8   MEANS   2.3X - 5.3Y = 4.8

HOW MANY PROBLEMS?2

TYPE 9E9 WHEN YOU NEED THE CALCULATOR.

PROBLEM 1.  SOLVE:

4.6  7.8  3.1
3.2  3.7  2.4

Y =?
```

At this point the student has a choice of answering or asking for the calculator.

Let's continue our RUN showing the work of a student who uses the calculator. Many students will find it helpful to make notes while running the programs.

```
Y =?9E9
TO USE THE CALCULATOR, FOLLOW THIS CODE:
ADDITION, 1; MULTIPLICATION, 2; DIVISION, 3; FINISHED, 0
INPUT CODE?3
DIVIDEND, DIVISOR:?7.8,4.6
QUOTIENT = 1.69565
INPUT CODE?3
DIVIDEND, DIVISOR:?3.1,4.6
QUOTIENT = .673913
INPUT CODE?2
2 FACTORS?-3.2,1.69565
PRODUCT =-5.42608
INPUT CODE?2
2 FACTORS?-3.2,.673913
PRODUCT =-2.15652
INPUT CODE?1
2 ADDENDS?-5.42608,3.7
SUM = -1.72608
INPUT CODE?1
2 ADDENDS?-2.15652,2.4
SUM = .24348
INPUT CODE?3
DIVIDEND, DIVISOR:?.24345,-1.72608
QUOTIENT =-.141042
INPUT CODE?0

Y =?-.141
VERY GOOD!!! (10 POINTS) COMPUTER HAD Y =-.141058

AND X =?9E9
TO USE THE CALCULATOR, FOLLOW THIS CODE:
ADDITION, 1; MULTIPLICATION, 2; DIVISION, 3; FINISHED, 0
INPUT CODE?2
2 FACTORS?7.8,-.141
PRODUCT =-1.0998
INPUT CODE?1
2 ADDENDS?3.1,1.0998
SUM = 4.1998
INPUT CODE?3
DIVIDENT, DIVISOR:?4.1998,4.6
QUOTIENT = .913
INPUT CODE?0

AND X =?.913
CORRECT!!! (5 POINTS)  COMPUTER HAD X = .913099

PROBLEM 2.  SOLVE:

 4.6   -5.5   -1.1
-8      5.3   -5.5

Y =?
```

Divide both members of first equation by 4.6 obtaining:

1	1.69565	.673913
3.2	3.7	2.4

Multiply both members of first equation by -3.2, obtaining:

-3.2	-5.42608	-2.15652
3.2	3.7	2.4

Add the two equations, obtaining:

0 $\quad -1.72608 \quad$ .24348

$.24348 \div (-1.72608) = -1.41042$

Difference caused by round-off errors.

$4.6x + (7.8)(-.141) = 3.1$

$4.6x + (-1.0998) = 3.1$

$4.6x = 3.1 + 1.0998$

$4.6x = 4.1998$

$x = 4.1998 \div 4.6 = .913$

Difference caused by round-off errors.

Unit 6C. Computer Solution of Linear Systems

There are several methods for solving systems of linear equations. An efficient method to use with a computer is called Gaussian elimination (named after the German mathematician, Carl Friedrich Gauss).

To illustrate this method, we show the solution of:

$$3x - 6y = 9$$
$$2x - 2y = 8$$

1. Divide the coefficients of the *first* equation by the coefficient of x in that equation (which is 3 in our example).

   3 −6 9 ⇒ 1 −2 3
   2 −2 8 ⇒ 2 −2 8

2. Multiply the coefficients of the *new first* equation by the coefficient of x in the second equation (2 in our example).

   1 −2 3 ⇒ 2 −4 6
   2 −2 8 ⇒ 2 −2 8

3. Subtract the coefficients of the *new first* equation from the coefficients of the *second* equation.

   2 −4 6 ⇒ 2 −4 6
   2 −2 8 ⇒ 0 2 2

4. This results in a *zero* coefficient of x in the *new second* equation. This can now be solved for y.

   0 2 2 means $2y = 2$
   $y = 1$

5. Using this value of y, solve the *original first* equation for x.

   $3x - 6(1) = 9$
   $x = (9 + 6)/3 = 5$

NOTE: This method is essentially the same as that used in Units 6A and 6B. There, instead of multiplying and subtracting as in steps 2 and 3 above, we multiplied by the negative and added.

6C-1: GAUSS-2 Write a program that accepts the coefficients of two linear equations as input, and which solves for X and Y using the Gaussian elimination method.

In order to generalize this method for more equations in more variables, it is useful to use subscripts. For example, three equations in three variables can be written as:

$$a_{11}x_1 + a_{12}x_2 + a_{13}x_3 = a_{14}$$
$$a_{21}x_1 + a_{22}x_2 + a_{23}x_3 = a_{24}$$
$$a_{31}x_1 + a_{32}x_2 + a_{33}x_3 = a_{34}$$

6C-2: GAUSS-3 Expand program 6C-1 to handle 3 equations in 3 variables. Here's a sample RUN.

```
RUN


LINEAR SYSTEMS IN 3 VARIABLES

INPUT THE COEFFICIENTS OF EQUATION 1
?1,2,3,26
INPUT THE COEFFICIENTS OF EQUATION 2
?4,3,7,59
INPUT THE COEFFICIENTS OF EQUATION 3
?5,8,7,82
FIRST REDUCTION
 1  2  3  26
 0 -5 -5 -45
 0 -2 -8 -48
SECOND REDUCTION
 1  2  3  26
 0  1  1  9
 0  0 -6 -30
THE SOLUTION IS
X(3) = 5
X(2) = 4
X(1) = 3
```

Here's a listing of a program to solve 3 equations in 3 variables using the Gaussian elimination method, as shown in the sample RUN above.

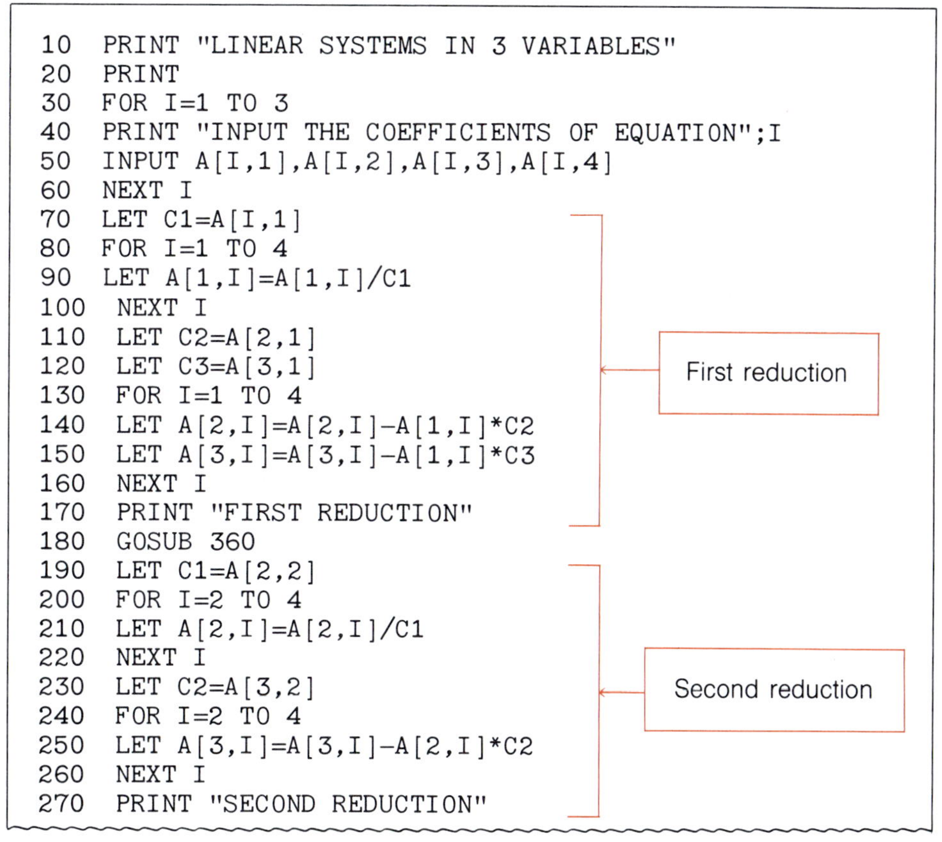

```
10  PRINT "LINEAR SYSTEMS IN 3 VARIABLES"
20  PRINT
30  FOR I=1 TO 3
40  PRINT "INPUT THE COEFFICIENTS OF EQUATION";I
50  INPUT A[I,1],A[I,2],A[I,3],A[I,4]
60  NEXT I
70  LET C1=A[I,1]
80  FOR I=1 TO 4
90  LET A[1,I]=A[1,I]/C1
100  NEXT I
110  LET C2=A[2,1]
120  LET C3=A[3,1]
130  FOR I=1 TO 4
140  LET A[2,I]=A[2,I]-A[1,I]*C2
150  LET A[3,I]=A[3,I]-A[1,I]*C3
160  NEXT I
170  PRINT "FIRST REDUCTION"
180  GOSUB 360
190  LET C1=A[2,2]
200  FOR I=2 TO 4
210  LET A[2,I]=A[2,I]/C1
220  NEXT I
230  LET C2=A[3,2]
240  FOR I=2 TO 4
250  LET A[3,I]=A[3,I]-A[2,I]*C2
260  NEXT I
270  PRINT "SECOND REDUCTION"
```

```
280  GOSUB 360
290  PRINT "THE SOLUTION IS"
300  LET X[3]=A[3,4]/A[3,3]
310  PRINT "X(3) =";X[3]
320  LET X[2]=A[2,4]-A[2,3]*X[3]
330  PRINT "X(2) =";X[2]
340  PRINT "X(1) =";A[1,4]-A[1,3]*X[3]-A[1,2]*X[2]
350  STOP
360  FOR I=1 TO 3
370  FOR J=1 TO 4
380  PRINT A[I,J];" ";
390  NEXT J
400  PRINT
410  NEXT I
420  RETURN
430  END
```

This subroutine prints the arrays.

6C-3: GAUSS-N Extend program 6C-2 to handle N equations in N variables, where N is an integer between 2 and 10 (specified by the person at the beginning of the RUN). Test your program with this example:

$$\begin{aligned} x_1 + 2x_2 - 12x_3 + 8x_4 &= 27 \\ 5x_1 + 4x_2 + 7x_3 - 2x_4 &= 4 \\ -3x_1 + 7x_2 + 9x_3 + 5x_4 &= 11 \\ 6x_1 - 12x_2 - 8x_3 + 3x_4 &= 49 \end{aligned}$$

Answer: (3, −2, 1, 5)

NOTE: An alternative method available on some systems is described in the Appendix.

Unit 6D. Pirate Rendezvous: Part I

Units 6D and 6E go together. Writing programs 6D-1 and 6D-2 will help you understand program 6E-1.

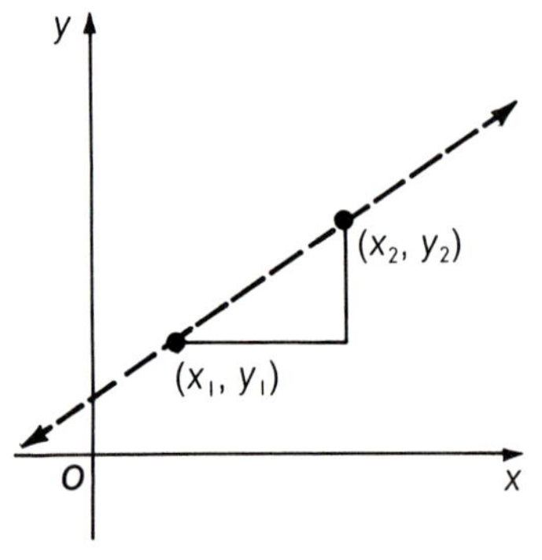

6D-1: PIRATE-PLOT Write a program that graphs the equation of a straight line which goes through *two given points*. You can modify program 5D-2 by computing the slope and y-intercept of a line from the coordinates of two points. The slope is:

$$m = \frac{y_2 - y_1}{x_2 - x_1}$$

To find the y-intercept, substitute in

$$y = mx + b$$

obtaining

$$b = y_1 - \left(\frac{y_2 - y_1}{x_2 - x_1}\right)x_1 \qquad \text{or} \qquad b = \frac{x_2y_1 - x_1y_2}{x_2 - x_1}$$

If you now replace statements 10 to 25 in program 5D-2 with statements which INPUT X1, Y1, X2, and Y2, and then compute M and B using the formulas given above, you should be all set.

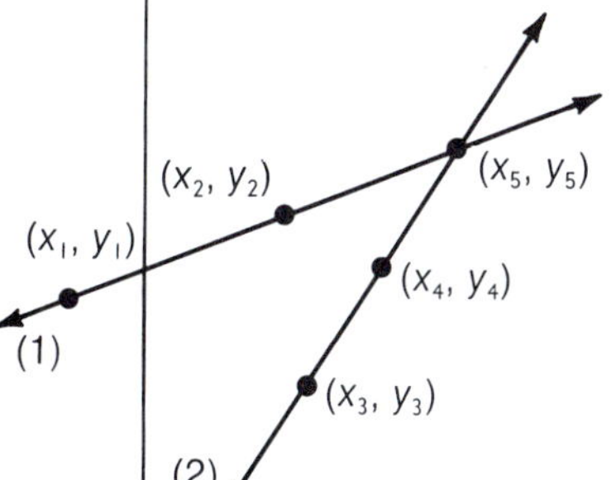

Now that we know how to graph the equation of a straight line through two given points, let's ask if we can find the point of intersection of *two* such straight lines. That is, given

$$(x_1, y_1), \qquad (x_2, y_2), \qquad (x_3, y_3), \qquad (x_4, y_4),$$

to find the coordinates

$$(x_5, y_5)$$

of their point of intersection.

Clearly, one way to do this is to write the equations in the form

$$(y_2 - y_1)x - (x_2 - x_1)y = x_1y_2 - x_2y_1 \qquad (1)$$
$$(y_4 - y_3)x - (x_4 - x_3)y = x_3y_4 - x_4y_3 \qquad (2)$$

and use the methods of Section 5. However, an alternative method is suggested in program 6D-2 for use in program 6E-1.

6D-2: RENDEZVOUS Write a program to find the coordinates (x_5, y_5) of the point of intersection of two straight lines when the coordinates (x_1, y_1) and (x_2, y_2) of two points on one and the coordinates (x_3, y_3) and (x_4, y_4) of two points on the other are given. Use:

$$y = m_1x + b_1 \qquad \text{where: } m_1 = (y_2 - y_1)/(x_2 - x_1)$$
$$b_1 = (x_2y_1 - x_1y_2)/(x_2 - x_1)$$

$$y = m_2x + b_1 \qquad \text{where: } m_2 = (y_4 - y_3)/(x_4 - x_3)$$
$$b_2 = (x_4y_3 - x_3y_4)/(x_4 - x_3)$$

The intersection of these lines is (x_5, y_5) where:

$$x_5 = (b_2 - b_1)/(m_1 - m_2), \qquad y_5 = (b_2m_1 - b_1m_2)/(m_1 - m_2)$$

Unit 6E. Pirate Rendezvous: Part II

Now let's have some fun with an anachronism—mixing pirate ships, radar, and computers.

6E-1: RADAR Radar has sighted two ships on straight-line courses for a rendezvous. It is suspected that ship 1 belongs to the dreaded Captain Skull, and that ship 2 is his supply ship. Question: Where do they intend meeting to transfer supplies and conduct other nefarious business? Given two sightings for each ship, compute the coordinates of the point of intersection, and graph both paths. Here is a sample RUN.

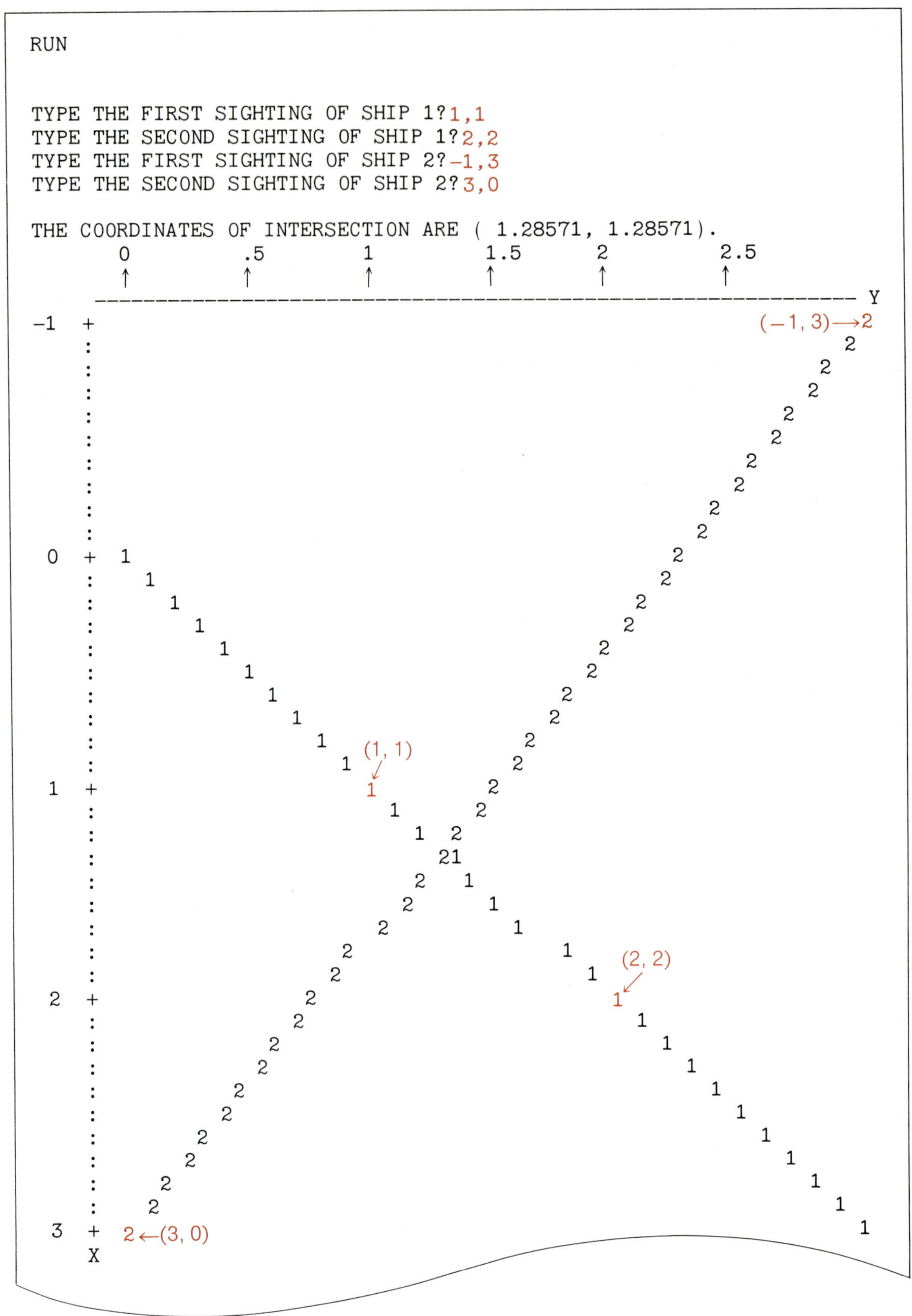

Here is a listing of a solution to program 6E-1.

```
10  PRINT "TYPE THE FIRST SIGHTING OF SHIP 1";
15  INPUT X[1],Y[1]
20  PRINT "TYPE THE SECOND SIGHTING OF SHIP 1";
25  INPUT X[2],Y[2]
30  PRINT "TYPE THE FIRST SIGHTING OF SHIP 2";
35  INPUT X[3],Y[3]
40  PRINT "TYPE THE SECOND SIGHTING OF SHIP 2";
45  INPUT X[4],Y[4]
50  LET M1=(Y[2]-Y[1])/(X[2]-X[1])
55  LET B1=(X[2]*Y[1]-X[1]-X[1]*Y[2])/(X[2]-X[1])
60  LET M2=(Y[4]-Y[3])/(X[4]-X[3])
65  Let B2=(X[4]*Y[3]-X[3]*Y[4])/(X[4]-X[3])
70  LET X[5]=(B2-B1)/(M1-M2)
75  LET Y[5]=(B2*M1-B1*M2)/(M1-M2)
80  PRINT
85  PRINT "THE COORDINATES OF INTERSECTION ARE (";X[5];",";Y[5];")."
90  PRINT
95  LET D1=9999
100  LET D2=-9999
105  LET R1=9999
110  LET R2=-9999
115  FOR I=1 TO 5
120  IF D1 <= X[I] THEN 130
125  LET D1=X[I]
130  IF D2 >= X[I] THEN 140
135  LET D2=X[I]
140  IF R1 <= Y[I] THEN 150
145  LET R1=Y[I]
150  IF R2 >= Y[I] THEN 160
155  LET R2=Y[I]
160  NEXT I
165  LET K1=(41-1)/(D2-D1)
170  LET K2=(61-1)/(R2-R1)
175  LET J1=21-K1*(D2+D1)/2
180  LET J2=31-K2*(R2+R1)/2
185  LET X1=X[1]*K1+J1
190  LET X2=X[2]*K1+J1
195  LET Y1=Y[1]*K2+J2
200  LET Y2=Y[2]*K2+J2
205  LET X3=X[3]*K1+J1
210  LET X4=X[4]*K1+J1
215  LET Y3=Y[3]*K2+J2
220  LET Y4=Y[4]*K2+J2
225  FOR I=1 TO 60 STEP 10
230  PRINT TAB(5+I);(I-J2)/K2;
235  NEXT I
240  PRINT
245  FOR I=1 TO 60 STEP 10
250  PRINT TAB(6+I);"+";
255  NEXT I
260  PRINT
265  PRINT TAB(5);"--";
270  FOR I=1 TO 61
275  PRINT "-";
280  NEXT I
285  PRINT " Y"
```

(Lines 95–160) These steps determine the domain and range so that the four given points and the point of intersection will all be on the graph.

(Lines 165–170) These steps determine the scales.

(Lines 175–180) These steps determine the position of the origin.

(Lines 185–220) These steps set up the corresponding coordinates in terms of K1, J1, K2, J2.

(Lines 225–285) These loops print the horizontal scale.

```
290  FOR I=1 TO 41
295  IF INT((I-1)/10 <> (I-1)/10 THEN 310
300  PRINT INT(10*(I-J1)/K1+.5)/10;TAB(4);"+";
305  GOTO 315
310  PRINT TAB(4);":";
315  LET X=I
320  LET Z1=((Y2-Y1)*X+X2*Y1-X1*Y2)/(X2-X1)
325  LET Z2=((Y4-Y3)*X+X4*Y3-X3*Y4)/(X4-X3)
330  IF Z1>Z2 THEN 370
335  IF Z1<1 THEN 350
340  IF Z1>61 THEN 400
345  PRINT TAB(6+Z1+.5);"1";
350  IF Z2<1 THEN 400
355  IF Z2>61 THEN 400
360  PRINT TAB(6+Z2+.5);"2";
365  GOTO 400
370  IF Z2<1 THEN 385
375  IF Z2>61 THEN 400
380  PRINT TAB(6+Z2+.5);"2";
385  IF Z1<1 THEN 400
390  IF Z1>61 THEN 400
395  PRINT TAB(6+Z1+.5);"1";
400  PRINT
405  NEXT I
410  PRINT "     X"
415  END
```

Z1 is the value used in graphing line 1, and Z2 is the value used in graphing line 2. Compare equations (1) and (2) on page 113.

This loop prints the vertical scale and the graph.

RUN this program several times using other coordinates. For example, try:

(0, −1), (3, −2), (1, −1), (2, −2)
(1, 1), (1.5, 1.5), (1, 0.5), (2, 1.75)

NOTE: This program would be much more realistic if the output could be displayed on a high-speed cathode-ray terminal to give an instantaneous picture of the predicted paths of the two ships.

6E-2: TIME-PREDICT Modify program 6E-1 so that you can input the *time* of each sighting. Then, after graphing the paths of the ships, the program should print the predicted time of arrival of each ship at the rendezvous.

HINTS:
Distance between two points = SQR((X2−X1)↑2+(Y2−Y1)↑2)

Read over Unit 3C.

SECTION 7 QUADRATIC EQUATIONS AND FUNCTIONS

Checklist of Computing Skills

Previously explained	PRINT, END, LET, INPUT, IF . . . THEN, GOTO, FOR, NEXT, STEP, SQR, INT, PRINT TAB, Subscripted variables—single subscripts, RND
Requires extra reading	String variables (Program 7E-3)
Special technique	Correcting binary conversion errors in integers (shown in program 7E-1)

Checklist of Algebraic Skills

1. $(3x + 4)(5x - 2) =$ __?__
2. Two linear factors of $8x^2 + 10x + 3$ are __?__ and __?__.
3. Use the quadratic formula to find the roots of $2x^2 + 5x - 12 = 0$: {__?__, __?__}
4. Sketch the graph of $y = x^2 - 4$.

   Check your answers with those printed at the bottom of the page.

Unit 7A. EXAM-CRAM, Part I

The final eight coaching units (7A and 7B, 8A and 8B, 9A and 9B, 10A and 10B) have one purpose—to help you and your classmates build a large practice program that will help students get ready for a final examination in algebra. Try out all these programs and make use of students' suggestions for improving them. Your teacher may suggest other topics to be included in EXAM-CRAM. If your EXAM-CRAM gets large enough, and if the data are generated randomly, you might even use it to generate problems for your final.

7A-1: PRODB Write a program to drill a student in multiplying two binomials. Generate four positive coefficients randomly and check the student's answer. In the sample RUN below, P(1), P(2), and P(3) denote the three coefficients of the product.

```
RUN


THIS SECTION TESTS YOUR ABILITY TO MULTIPLY BINOMIALS.
FOR EACH PRODUCT INPUT THE COEFFICIENTS FROM LEFT TO RIGHT.

( 7X + 8)( 6X + 6) =
P(1) =?42
P(2) =?90
```

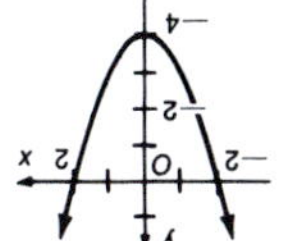

Answers: (1) $15x^2 + 14x - 8$ (2) $4x + 3$, $2x + 1$ (3) $\{-4, 1.5\}$ (4)

```
P(3) =?48
I GOT 42X↑2 + 90X + 48 TOO.

WOULD YOU LIKE TO DO ANOTHER (1=YES, 0=NO)?1

( 6X + 6)( 7X + 2) =
P(1) =?42
P(2) =?54
P(3) =?12
I GOT 42X↑2 + 54X + 12 TOO.

WOULD YOU LIKE TO DO ANOTHER (1=YES, 0=NO)?0
```

Here is a sample listing:

```
1  RANDOMIZE     (Use only if necessary.)
40  PRINT
45  PRINT "THIS SECTION TESTS YOUR ABILITY TO MULTIPLY BINOMIALS."
50  PRINT "FOR EACH PRODUCT INPUT THE COEFFICIENTS FROM LEFT TO RIGHT."
55  PRINT
60  LET C[1]=INT(9*RND(1)+1)
65  LET C[2]=INT(10*RND(1))
70  LET C[3]=INT(9*RND(1)+1)
75  LET C[4]=INT(10*RND(1))
80  PRINT "(";C[1];"X +";C[2];")(";C[3];"X +";C[4];") ="
85  PRINT "P(1) =";
90  INPUT P[1]
95  PRINT "P(2) =";
100  INPUT P[2]
105  PRINT "P(3) =";
110  INPUT P[3]
115  LET A[1]=C[1]*C[3]
120  LET A[2]=C[1]*C[4]+C[2]*C[3]
125  LET A[3]=C[2]*C[4]
130  PRINT "I GOT";A[1];"X↑2 +";A[2];"X +";A[3];
135  IF P[1] <> A[1] THEN 160
140  IF P[2] <> A[2] THEN 160
145  IF P[3] <> A[3] THEN 160
150  PRINT " TOO."
155  GOTO 165
160  PRINT ". CHECK YOUR WORK."
165  PRINT
170  PRINT "WOULD YOU LIKE TO DO ANOTHER (1=YES, 0=NO)";
175  INPUT X
180  IF X=1 THEN 55
185  END
```

7A-2: PRODBN Modify program 7A-1 so that coefficients C(2) and C(4) may be negative.

7A-3: PRODBNH Modify program 7A-2 to check each coefficient as the student types it in.

7A-4: FACTOR Write a program to drill a student in factoring trinomials. Below is a sample RUN and a listing. Have several students try it, and make use of their criticisms to improve the program.

```
RUN



THIS SECTION TESTS YOUR ABILITY TO FACTOR TRINOMIALS.
YOU WILL BE GIVEN A TRINOMIAL AND THEN ASKED FOR ITS
TWO FACTORS, ONE COEFFICIENT AT A TIME.

FIND THE FACTORS OF:
 20X↑2 + 24X + 4
TYPE THE X COEFFICIENT OF FACTOR 1?4
TYPE THE CONSTANT TERM OF THIS FACTOR?4
THAT'S RIGHT
TYPE THE X COEFFICIENT OF FACTOR 2?5
TYPE THE CONSTANT TERM OF THIS FACTOR?1
THAT'S RIGHT
( 5X + 1) AND ( 4X + 4)

WOULD YOU LIKE TO DO ANOTHER (1=YES, 0=NO)?1

FIND THE FACTORS OF:
 15X↑2 + 26X + 8
TYPE THE X COEFFICIENT OF FACTOR 1?5
TYPE THE CONSTANT TERM OF THIS FACTOR?4
NO; THAT IS NOT CORRECT.
IF YOU WOULD LIKE TO TRY AGAIN, TYPE 1; IF YOU WANT
A HINT, TYPE 0:?0
ONE OF THE FACTORS IS ( 3X + 4)
TYPE THE X COEFFICIENT OF FACTOR 2?5
TYPE THE CONSTANT TERM OF THIS FACTOR?2
THAT'S RIGHT
( 3X + 4) AND ( 5X + 2)

WOULD YOU LIKE TO DO ANOTHER (1=YES, 0=NO)?1

FIND THE FACTORS OF:
 12X↑2 + 27X + 15
TYPE THE X COEFFICIENT OF FACTOR 1?4
TYPE THE CONSTANT TERM OF THIS FACTOR?5
THAT'S RIGHT
TYPE THE X COEFFICIENT OF FACTOR 2?6
TYPE THE CONSTANT TERM OF THIS FACTOR?3
NO; THAT IS NOT CORRECT.
THE CORRECT FACTORS ARE:
( 4X + 5) AND ( 3X + 3)

WOULD YOU LIKE TO DO ANOTHER (1=YES, 0=NO)?0
```

Here is a sample listing.

```
1  RANDOMIZE    (Use only if necessary.)
200  PRINT
205  PRINT "THIS SECTION TESTS YOUR ABILITY TO FACTOR TRINOMIALS."
210  PRINT "YOU WILL BE GIVEN A TRINOMIAL AND THEN ASKED FOR ITS"
215  PRINT "TWO FACTORS, ONE COEFFICIENT AT A TIME."
220  PRINT
225  FOR I=1 TO 4
230  LET C[I]=INT(5*RND(1)+1)
235  NEXT I
240  PRINT "FIND THE FACTORS OF:"
245  PRINT C[1]*C[3];"X↑2 +";C[1]*C[4]+C[2]*C[3];"X +";C[2]*C[4]
250  LET I=1
255  PRINT "TYPE THE X COEFFICIENT OF FACTOR";I;
260  INPUT A1
265  PRINT "TYPE THE CONSTANT TERM OF THIS FACTOR";
270  INPUT A2
275  IF A1 <> C[1] THEN 325
280  IF A2 <> C[2] THEN 325
285  LET D[1]=C[1]
290  LET C[1]=0
295  LET D[2]=C[2]
300  LET C[2]=0
305  PRINT "THAT'S RIGHT"
310  IF I <> 1 THEN 405
315  LET I=I+1
320  GOTO 255
325  IF A1 <> C[3] THEN 360
330  IF A2 <> C[4] THEN 360
335  LET D[3]=C[3]
340  LET C[3]=0
345  LET D[4]=C[4]
350  LET C[4]=0
355  GOTO 305
360  PRINT "NO; THAT IS NOT CORRECT."
365  IF I=2 THEN 400
370  PRINT "IF YOU WOULD LIKE TO TRY AGAIN, TYPE 1; IF YOU WANT"
375  PRINT "A HINT, TYPE 0:";
380  INPUT X
385  IF X=1 THEN 255
390  PRINT "ONE OF THE FACTORS IS (";C[1];"X +";C[2];")"
395  GOTO 315
400  PRINT "THE CORRECT FACTORS ARE:"
405  FOR I=1 TO 4
410  IF C[I] <> 0 THEN 420
415  LET C[I]=D[I]
420  NEXT I
425  PRINT "(";C[1];"X +";C[2];") AND (";C[3];"X +";C[4];")"
430  PRINT
435  PRINT "WOULD YOU LIKE TO DO ANOTHER (1=YES, 0=NO)";
440  INPUT X
445  IF X=1 THEN 220
450  END
```

7A-5: FACTORN Modify program 7A-4 so that constant terms in the factors may be negative.

Unit 7B. EXAM-CRAM, Part II

In the B units of EXAM-CRAM we will not give any sample runs or listings—just some further ideas for programs.

7B-1: CLUE Modify 7A-1 so that if a student gives a wrong answer, a hint will be given, with a second chance to answer, before the correct answer is given.

7B-2: MOTION-PROB Write a program to drill a student in solving a motion problem. Here's an example:

> A girl walked to a bicycle shop at the rate of <u>2</u> Km per hour, and then rode a bicycle back the same distance at the rate of <u>10</u> Km per hour. If the entire trip took <u>3</u> hours, how far and how long did she walk?

The underlined numbers should be generated randomly within suitable limits.

Unit 7C. The Quadratic-Equation-Solving Machine

A computer program is something like a machine that accepts input as its raw material and produces output as its finished product. A good example of this is a program that uses the quadratic formulas:

R1 = (−B+SQR(B*B−4*A*C))/2*A

and

R2 = (−B−SQR(B*B−4*A*C))/2*A

7C-1: QUADSOLVE-1 Write a program which uses the quadratic formulas to find the two roots of a quadratic equation. The input to your program should be the coefficients A, B, and C. The output should be the roots R1 and R2. Your program should test for special difficulties and print appropriate messages

(*Hint:* There will be difficulty if your program tries to divide by zero, or if it tries to take the square root of a negative number). A sample RUN might look like this:

```
RUN


QUADRATIC EQUATION SOLVER:

WHAT ARE A, B, C?1,2,3
NO REAL ROOTS

WHAT ARE A, B, C?0,2,3
NOT A QUADRATIC EQUATION

WHAT ARE A, B, C?1,-3,2
R1 = 2
R2 = 1

WHAT ARE A, B, C?
```

7C-2: QUADSOLVE-2 Modify the preceding program so that the roots are given in "radical form" when B^2-4AC is not a perfect square.

Unit 7D. Quadratic Graphs

In this unit we'll study several ways of graphing quadratic functions. First let's look at a program that is similar in some respects to program 5D-1 for a straight-line graph.

7D-1: QUAD-GRAPH1 Write a program that computes the range for a given domain and then graphs the function. Here's a sample solution for graphing the function defined by $y = -x^2 + 12x$.

```
10  LET A=1
20  LET B=11
30  LET R1=9999
40  LET R2=-9999
50  FOR X=A TO B
60  LET Y=-X↑2+12*X
70  IF R1 <= Y THEN 90
80  LET R1=Y
90  IF R2 >= Y THEN 110
100  LET R2=Y
110  NEXT X
120  IF 2*(R2-R1)>56 THEN 280
130  FOR I=R1 TO R2+2 STEP 2
140  PRINT TAB(7+2*(I-R1));I;
150  NEXT I
160  PRINT
170  PRINT TAB(5);"---+";
180  FOR I=1 TO (R2+2-R1)/2
```

These steps determine the domain. (lines 10–20)

These steps compute the range. (lines 30–110)

The loops in lines 130–210 print the horizontal scale.

```
190  PRINT "---+";
200  NEXT I
210  PRINT "---Y"
220  FOR X=A TO B
230  LET Y=-X↑2+12*X
240  PRINT X;TAB(8+2*(Y-R1));"*"
250  NEXT X
260  PRINT " X"
270  STOP
280  PRINT "TOO WIDE--R1 =";R1;" AND R2 =";R2
290  END
RUN
```

This loop prints the vertical scale and the graph.

```
      11  13  15  17  19  21  23  25  27  29  31  33  35  37
   ---+---+---+---+---+---+---+---+---+---+---+---+---+---+---Y
1     *
2                       *
3                                     *
4                                               *
5                                                     *
6                                                       *
7                                                     *
8                                               *
9                                     *
10                      *
11    *
X

END
60 LET Y=X↑2
230 LET Y=X↑2
RUN
```

We'll try another function.

```
TOO WIDE--R1 = 1 AND R2 = 121

END
10 LET A=-5
20 LET B=5
RUN
```

We'll change the domain.

```
      0   2   4   6   8   10  12  14  16  18  20  22  24  26
   ---+---+---+---+---+---+---+---+---+---+---+---+---+---+---Y
-5                                                      *
-4                                    *
-3                      *
-2            *
-1      *
 0    *
 1      *
 2            *
 3                      *
 4                                    *
 5                                                      *
 X

END
```

Another way to plan a graph that will be a portion of a parabola is to find the coordinates of the vertex first. Let's see how this can be done:

$y = ax^2$
Vertex: $(0, 0)$

The computation for locating the vertex of the graph of

$$y = ax^2 + bx + c$$

is as follows: We write

$$y = a\left(x^2 + \frac{b}{a}x\right) + c = a\left(x^2 + \frac{bx}{a} + \frac{b^2}{4a^2}\right) + c - \frac{b^2}{4a},$$

from which we have

$$y - \left(c - \frac{b^2}{4a}\right) = a\left(x + \frac{b}{2a}\right)^2,$$

and the coordinates of the vertex are $\left(-\frac{b}{2a}, c - \frac{b^2}{4a}\right)$, or

$$\left(-\frac{b}{2a}, -\frac{b^2 - 4ac}{4a}\right).$$

This tells us that the parabola is symmetric about the line $x = -\frac{b}{2a}$.

$y = ax^2 + c$
Vertex: $(0, c)$

NOTE: If $b^2 - 4ac = 0$, the coordinates of the vertex are $(-\frac{b}{2a}, 0)$, the parabola is tangent to the x-axis, and there is only one real root.

If $b^2 - 4ac < 0$, then the parabola does not intersect the x-axis and there are no real roots.

If $b^2 - 4ac > 0$, then the parabola intersects the x-axis, and there are two real roots.

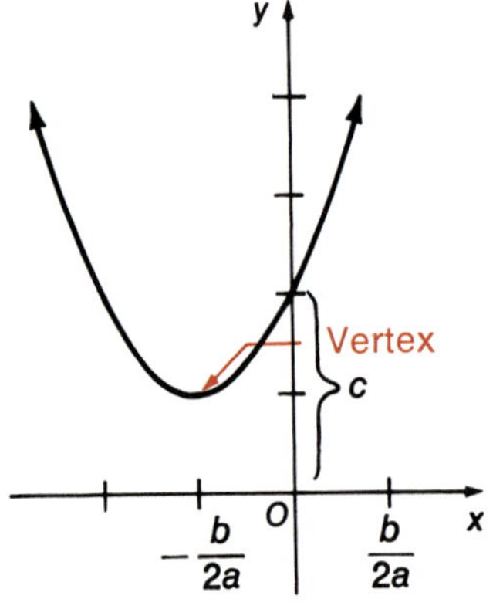

$y = ax^2 + bx + c$
Vertex: $\left(-\frac{b}{2a}, -\frac{b^2 - 4ac}{4a}\right)$

7D-2: QUAD-GRAPH2 Write a program to graph a portion of the function defined by:

$$y = 0.1x^2 - 0.2x$$

Using the formulas for the coordinates of the vertex, we find that the vertex is:

$$(1, -0.1)$$

This suggests a domain for x as:

$$D = \{-1 \le x \le 3\}$$

We compute $(-1, 0.3)$ and $(3, 0.3)$ and find that the range of y will be:

$$R = \{-0.1 \le y \le 0.3\}$$

We shall take 10 line-feeds to make one vertical unit, and we shall expand the Y-scale so that 16 spaces represent 0.1. These choices are represented in the program by the following steps:

1. To spread the X-scale out, we'll use a loop FOR I = −10 TO 30 to control the *number* of times we graph a value of X, but we'll make X = I/10 when it comes time to calculate Y as a function of X.

2. To move the X-axis far enough to the right (so that we can graph points where Y is negative), we'll shift everything 18 spaces to the right.

3. To spread the Y-scale, we'll multiply the value of Y by a scale factor of 160.

Look at the sample RUN below.

```
RUN              -1                                               *
                 -.9                                        *
                 -.8                                  *
                 -.7                            *
                 -.6                       *
                 -.5                  *
                 -.4             *
                 -.3         *
                 -.2     *
                 -.1 *
-.1-------↑-------0-------↑------.1-------↑------.2-------↑------.3-- Y
               *  .1
            *     .2
          *       .3
        *         .4
      *           .5
     *            .6
   *              .7
   *              .8
  *               .9
  *               1
  *               1.1
   *              1.2
   *              1.3
     *            1.4
      *           1.5
        *         1.6
          *       1.7
            *     1.8
               *  1.9
                  *
                  2.1*
                  2.2    *
                  2.3        *
                  2.4            *
                  2.5                 *
                  2.6                      *
                  2.7                           *
                  2.8                                 *
                  2.9                                       *
                  3                                              *
```

Here is a listing of the program that produced the run on page 125.

```
10  FOR I=-10 TO 30
20  IF I <> 0 THEN 60
30  PRINT "-.1-------↑-------0-------↑------.1";
40  PRINT "-------↑------.2-------↑------.3-- Y"
50  GOTO 160
60  LET X=I/10
70  LET Y=.1*X↑2-.2*X
80  LET Y=Y*160+18
90  IF Y <> 18 THEN 120
100  PRINT TAB(18);"*"
110  GOTO 160
120  IF Y<18 THEN 150
130  PRINT TAB(17);X;TAB(Y+.5);"*"
140  GOTO 160
150  PRINT TAB(Y+.5);"*";TAB(17);X
160  NEXT I
170  END
```

Make these changes

```
10  FOR 1=-30 TO 30
82  IF Y>68 THEN 160
84  IF Y<0 THEN 160
```

and then try:

```
(a) 70  LET Y = .1*X↑2
(b) 70  LET Y = -.1*X↑2
```

7D-3: SUPER-GRAPH If you try the preceding program on other functions, you will find that in some cases the * is displaced when the numbers on the X-axis get in the way. Can you improve the program so that in cases like this, the X number doesn't get printed, and the * appears instead?

Hint: Study lines 90 and 100; these handle part of the problem.

Unit 7E. The Quadratic Cipher

The simplest way to put a message into coded form is to substitute a new symbol (say a number) for each letter in the message. If, for example, we let A = 1, B = 2, C = 3, . . . , Z = 26, then the message

S E C R E T A G E N T

becomes

19, 5, 3, 18, 5, 20, 1, 7, 5, 14, 20.

A simple substitution scheme like this is called a *cipher.** This is not a very difficult cipher to break, since there are only 26 numbers involved, and symbols for letters that are repeated often in English (for example, "e") also repeat in the message (as "5" in our example).

7E-1: QUADCIPH Write a program that takes a message written in the above simple 26-number cipher, and then does a second encipherment. Let's call this second process "scrambling" (some books call such a process "superencipherment"). Let's also insist that:

(a) the scrambling be different every time the program is run,

(b) more than 26 numbers be generated, and

(c) the same letter (like "e") doesn't always get scrambled into the same number.

Then write a program to unscramble back to the original 26-number cipher.

The method we'll use can be illustrated by scrambling a few cipher numbers by hand. The first step is to randomly choose three positive "scrambling numbers," say C(1)=1, C(2)=2, C(3)=0. We'll then put these three numbers in front of our message as follows:

	—	—	—	S	E	C	R	E
	C(1)	C(2)	C(3)	C(4)	C(5)	C(6)	C(7)	C(8)
	1	2	0	19	5	3	18	5
S(1)	C(1)	C(2)	C(3)	X				
S(2)		C(2)	C(3)	C(4)	X			
S(3)			C(3)	C(4)	C(5)	X		
S(4)				C(4)	C(5)	C(6)	X	
S(5)					C(5)	C(6)	C(7)	X

Original message preceded by 3 scrambling numbers

*A cipher is a special case of what is more generally called a *code*.

We'll now calculate the first scrambled number, S(1), by substituting X=C(4)=19 in the quadratic scrambling polynomial:

| S |

$$\begin{aligned} S(1) &= C(1)*X*X + C(2)*X + C(3) \\ &= C(1)*C(4)*C(4) + C(2)*C(4) + C(3) \\ &= 1*19*19 + 2*19 + 0 = 399 \end{aligned}$$

Now let's make X = C(5) = 5, and use the three values of C(1) to the left of X in our quadratic scrambling polynomial:

| E |

$$\begin{aligned} S(2) &= C(2)*C(5)*C(5) + C(3)*C(5) + C(4) \\ &= 2*5*5 + 0*5 + 19 = 69 \end{aligned}$$

Continuing:

| C |

$$\begin{aligned} S(3) &= C(3)*C(6)*C(6) + C(4)*C(6) + C(5) \\ &= 0*3*3 + 19*3 + 5 = 62 \end{aligned}$$

| R |

$$\begin{aligned} S(4) &= C(4)*C(7)*C(7) + C(5)*C(7) + C(6) \\ &= 19*18*18 + 5*18 + 3 = 6249 \end{aligned}$$

| E |

$$\begin{aligned} S(5) &= C(5)*C(8)*C(8) + C(6)*C(8) + C(7) \\ &= 5*5*5 + 3*5 + 18 = 158 \end{aligned}$$

and so on. Thus, "E" is coded in two different ways (69 and 158) in our message!

To *unscramble* the message, we must solve a quadratic equation for each number. For example, to unscramble 399, we use the fact that the original cipher number was a number X such that:

$$1*X*X + 2*X + 0 = 399$$

or

$$1*X*X + 2*X - 399 = 0$$

Using the quadratic formula, we have:

$$X = (-2 \pm SQR(4 + 1596))/2 = -1 \pm 20 = 19 \text{ or } -21$$

We choose the value 19, since it lies in $\{1 \leq X \leq 26\}$. Actually, choosing the positive square root will give the right answer, since the coefficients of the scrambling polynomial are positive.

Now that we have the 19, we can unscramble the second number by solving:

$$2*X*X + 0*X + 19 = 69$$
$$X = (SQR(400))/4 = 5$$

The process is continued until we reach the end of the message.

The following program will scramble or unscramble a message of up to 30 letters.

```
5   RANDOMIZE    (Use only if necessary.)
10  DIM C[30],S[30]
20  PRINT "SCRAMBLE (1) OR UNSCRAMBLE (0)";
30  INPUT A
40  IF A=0 THEN 270
50  PRINT "HOW MANY NUMBERS TO BE SCRAMBLED";
60  INPUT N
70  LET C[1]=INT(5*RND(1))
80  LET C[2]=INT(5*RND(1))
90  LET C[3]=INT(5*RND(1))
100 FOR I=1 TO N
110 PRINT "#";I;
120 INPUT C[I+3]
130 NEXT I
140 FOR I=1 TO N
150 LET S[I]=C[I]*C[I+3]*C[I+3]+C[I+1]*C[I+3]+C[I+2]
160 NEXT I
170 PRINT "SCRAMBLED MESSAGE IS:"
180 FOR I=1 TO N-1
190 PRINT S[I];
200 NEXT I
210 PRINT S[N]
220 PRINT TAB(40);"*TOP SECRET*"
230 PRINT "YOUR SCRAMBLING COEFFICIENTS ARE:";TAB(41);
240 PRINT C[1];",";C[2];",";C[3]
250 PRINT "(MEMORIZE AND DESTROY)";TAB(40);"*TOP SECRET*"
260 STOP
270 PRINT "HOW MANY NUMBERS TO BE UNSCRAMBLED";
280 INPUT N
290 PRINT "INPUT THE THREE SCRAMBLING COEFFICIENTS";
300 INPUT C[1],C[2],C[3]
310 PRINT "INPUT SCRAMBLED NUMBERS, ONE AT A TIME:"
320 FOR I=1 TO N
330 PRINT "S(";I;")";
340 INPUT S[I]
350 NEXT I
360 FOR I=1 TO N
370 LET D=SQR(C[I+1]*C[I+1]-4*C[I]*(C[I+2]-S[I]))
380 LET C[I+3]=(-C[I+1]+D)/(2*C[I])
390 LET C[I+3]=INT(C[I+3]+.05)
400 NEXT I
410 PRINT "UNSCRAMBLED NUMBERS ARE:"
420 FOR I=1 TO N-1
430 PRINT C[I+3];
440 NEXT I
450 PRINT C[N+3]
460 END
```

NOTE: Because computers translate decimal numbers into binary before working on them (and vice versa before printing back answers), a little accuracy can be lost in the translation. Thus, a computer may get an answer of 12.9999 instead of 13. Normally this doesn't do any harm. If, however, we *know* that the answer is an integer, the trick in line 390 can be used to convert a decimal approximation back to the nearest integer.

Here is a RUN that scrambles:

```
RUN


SCRAMBLE (1) OR UNSCRAMBLE (0)?1
HOW MANY NUMBERS TO BE SCRAMBLED?27
# 1?19
# 2?5
# 3?3
# 4?18
# 5?5
# 6?20
# 7?1
# 8?7
# 9?5
# 10?14
# 11?20
# 12?19
# 13?13
# 14?5
# 15?5
# 16?20
# 17?1
# 18?20
# 19?20
# 20?5
# 21?14
# 22?15
# 23?3
# 24?12
# 25?15
# 26?3
# 27?11
SCRAMBLED MESSAGE IS:
1162 134 89 6249 158 1565 43 386 512 299 2914 2091 2645 608 545 5305
30 2401 8040 145 4205 4589 102 2199 3432 78 1620
                                             *TOP SECRET*
YOUR SCRAMBLING COEFFICIENTS ARE:              3, 4, 3
(MEMORIZE AND DESTROY)                       *TOP SECRET*
```

Here is a RUN that unscrambles:

```
RUN


SCRAMBLE (1) OR UNSCRAMBLE (0)?0
HOW MANY NUMBERS TO BE UNSCRAMBLED?27
INPUT THE THREE SCRAMBLING COEFFICIENTS?3,4,3
INPUT SCRAMBLED NUMBERS, ONE AT A TIME:
S( 1)?1162
S( 2)?134
S( 3)?89
S( 4)?6249
S( 5)?158
S( 6)?1565
```

```
S( 7)?43
S( 8)?386
S( 9)?512
S( 10)?299
S( 11)?2914
S( 12)?2091
S( 13)?2645
S( 14)?608
S( 15)?545
S( 16)?5305
S( 17)?30
S( 18)?2401
S( 19)?8040
S( 20)?145
S( 21)?4205
S( 22)?4589
S( 23)?102
S( 24)?2199
S( 25)?3432
S( 26)?78
S( 27)?1620
UNSCRAMBLED NUMBERS ARE:
 19 5 3 18 5 20 1 7 5 14 20 19 13 5 5 20 1 20 20 5 14 15 3 12 15 3 11
```

7E-2: ULTRA-CIPH A defect in our scrambling program is that letters at the beginning of the alphabet have lower numbers than those at the end of the alphabet, which might be a give-away to a decoding agent. See if you can rectify this defect.

7E-3: LETTER-CIPH **(a)** If your computer allows string manipulation, improve program 7C-1 so that messages written in English can be scrambled and unscrambled. **(b)** If your version of BASIC has statements that allow conversion of letters into their ASCII code,* use these to do the ciphering, and use program 7E-1 to do the scrambling.

If the subject of coding and decoding appeals to you, check your library for the book *The Code-Breakers* by David Kahn (New York: The Macmillan Co., 1967).

*This is the "American Standard Code for Information Interchange." The digits 0 to 9 have codes 48 to 57, the capital letters A to Z have codes 65 to 90. Other codes are used for symbols and lowercase letters.

SECTION 8 POLYNOMIAL EQUATIONS; RATIONAL EXPRESSIONS

Checklist of Computing Skills

Previously explained	PRINT, END, LET, INPUT, IF . . . THEN, STOP, GOTO, FOR, NEXT, SQR, ABS, INT, PRINT TAB Subscripted variables—single subscripts, RND
Explained in this section	DEF FNA(X)
Requires extra reading	String variables (Program 8A-2)
Special technique	Using nested form of a polynomial (Programs 8C-2 and 8C-3)

Checklist of Algebraic Skills

1. $\frac{2}{3} + \frac{4}{5} = \frac{?}{?}$
2. The GCF of 32 and 56 is ___?___.
3. The LCD of $\frac{5}{12}$ and $\frac{3}{16}$ is ___?___.
4. $\frac{2}{x+1} + \frac{3}{x-1} = \frac{?}{?}$
5. Evaluate $2x^2 + 3x + 4$ for each integer in the interval $-5 \le x \le 2$.
6. Using the values found in Exercise 5, give two intervals which contain roots of $2x^2 + 3x + 4 = 0$:
   (a) Between ___?___ and ___?___ (b) Between ___?___ and ___?___

Check your answers with those printed at the bottom of the page.

NEW: DEF FNA(X)

A special feature of BASIC allows us to DEFine a function *once* in a program, and then use it as often as we wish just by typing its name and argument. The name must use the two letters FN followed by a *single* letter of your choice. Thus, you can define up to 26 of your own functions in a program.

For example, it may be convenient to use a rounding function to round results to dollars and cents. Here is a RUN of a program to find average costs.

```
RUN


INPUT COSTS.  TYPE -1 TO END.
?2.95
?3.49
?3.23
?3.95
?3.39
?3.74
?-1
CALCULATED AV  ROUNDED AV
 3.45833         3.46
```

Answers: (1) 22, 15 (2) 8 (3) 48 (4) $5x + 1$, $x^2 - 1$ (5) 31, 16, 5, −2, −5, −4, 1, 10 (6) (a) −3, −2 (b) 0, 1

Here is a listing of the program:

```
10  LET S=0
20  LET N=0
30  DEF FNR(X)=INT(X*100+.5)/100
40  PRINT "INPUT COSTS.  TYPE -1 TO END."
50  INPUT C
60  IF C=-1 THEN 100
70  LET S=S+C
80  LET N=N+1
90  GOTO 50
100  LET A=S/N
110  PRINT "CALCULATED AV","ROUNDED AV"
120  PRINT A,FNR(A)
130  END
```

Here is where we define the function FNR.

Here is where we use FNR.

After a function has been DEFined, it can be used in PRINT statements, in IF statements, and on the right side of LET statements.

Unit 8A. EXAM-CRAM, Part III

8A-1: RAT Write a program that drills a student in adding rational numbers.

NOTE: The INPUT statement will not accept numbers written as fractions. This means that your program will have to ask for the numerator and denominator separately. (It is possible to get around this limitation if your computer accepts STRING INPUT.)

Here's a sample RUN:

```
RUN



THIS SECTION TESTS YOUR ABILITY TO ADD TWO RATIONAL
NUMBERS.  GIVE THE ANSWER AS 'NUMERATOR, DENOMINATOR'
AND DO NOT REDUCE TO LOWEST TERMS.

WHAT IS 6 / 5 +  5 / 4?49,20
CORRECT.  ANSWER:  49 / 20

WOULD YOU LIKE TO DO ANOTHER (1=YES, 0=NO)?1

WHAT IS  8 / 4 +  1 / 6?50,24
NO.  ANSWER:  52 / 24

WOULD YOU LIKE TO DO ANOTHER (1=YES, 0=NO)?0
```

8A-2: FRACT If your computer system can handle *string variables,* rewrite program 8A-1 to allow answers to be INPUT in the form 49/20.

8A-3: GCF Write a program which drills a student in finding the greatest common factor (GCF) of two positive integers. Get it to run and have several students try it out and suggest improvements.

```
RUN



THIS SECTION TESTS YOUR ABILITY TO FIND THE GREATEST
COMMON FACTOR (GCF) OF TWO POSITIVE INTEGERS.

WHAT IS THE GCF OF 74 AND 90?2
THAT'S RIGHT!

WOULD YOU LIKE TO DO ANOTHER (1=YES, 0=NO)?1

WHAT IS THE GCF OF 67 AND 69?1
THAT'S RIGHT!

WOULD YOU LIKE TO DO ANOTHER (1=YES, 0=NO)?1

WHAT IS THE GCF OF 66 AND 62?4
NO.  THE GCF IS: 2

WOULD YOU LIKE TO DO ANOTHER (1=YES, 0=NO)?0
```

8A-4: RATALG Write a program that gives practice in adding two simple algebraic rational expressions, for example:

$$\frac{3}{2x+4} + \frac{7}{x-3}$$

This is a difficult program. It would make a good project for a few interested students. Then the rest of the class could use it as practice. Programs 7A-1, 9A-1, and 9A-2 will give you some ideas.

Unit 8B. EXAM-CRAM, Part IV

8B-1: LCD Write a program that gives practice in finding the least common denominator (LCD) of two fractions.

8B-2: SMART-RAT Use the ideas in program 8A-3 to change program 8A-1 to give answers as follows:

NO; ANSWER IS: 10/24 = 5/12

8B-3: COMBO-1 Combine programs 7A-1 and 7A-4 into one longer program. By removing the END statement from 7A-1, a "NO" answer can be made to cause the program to drop through to the next higher line number which will be in 7A-4.

Unit 8C. Evaluating Polynomials

Anyone who has progressed this far in the book will find program 8C-1 easy. But it involves an important idea needed in units 8D and 9D. So let's make sure that you know how to handle the problem. In this unit, we'll also introduce two new programming techniques: "nested evaluation" in 8C-2 and "recursion" in 8C-3.

Here's a typical "polynomial evaluation" problem as given in an algebra book:

> Evaluate the function defined by $f(x) = 2x^2 + 3x - 4$ for x, where $-5 \leq x \leq 2$ and x is an integer.
>
> SOLUTION
>
> $x_1 = -5 \quad f(x_1) = 2(-5)^2 + 3(-5) - 4 = 31$
> $x_2 = -4 \quad f(x_2) = 2(-4)^2 + 3(-4) - 4 = 16$
> $x_3 = -3 \quad f(x_3) = 2(-3)^2 + 3(-3) - 4 = 5$
> $x_4 = -2 \quad f(x_4) = 2(-2)^2 + 3(-2) - 4 = -2$
> $x_5 = -1 \quad f(x_5) = 2(-1)^2 + 3(-1) - 4 = -5$
> $x_6 = 0 \quad f(x_6) = 2(0)^2 + 3(0) - 4 = -4$
> $x_7 = 1 \quad f(x_7) = 2(1)^2 + 3(1) - 4 = 1$
> $x_8 = 2 \quad f(x_8) = 2(2)^2 + 3(2) - 4 = 10$

8C-1: POLY-EVAL Do the above the easy way—on a computer.

```
10  PRINT " X"," F(X)"
20  FOR X=-5 TO 2
30  PRINT X,2*X*X+3*X-4
40  NEXT X
50  END
```

or

```
5  DEF FNA(X)=2*X*X+3*X-4
10  PRINT " X"," F(X)"
20  FOR X=-5 TO 2
30  PRINT X,FNA(X)
40  NEXT X
50  END
```

NESTED FORM OF A POLYNOMIAL

It turns out that evaluating a polynomial in the usual form in which we write it is not the most efficient way from the computer's point of view. For example, to evaluate

(1) F(X)=9*X↑8+8*X↑7+7*X↑6+6*X↑5+5*X↑4+4*X↑3+3*X↑2+2*X+1

takes about 44 arithmetic operations for each value of X (for example, 9*X↑8 involves 8 multiplications, 8*X↑7 takes 7, and so on; finally, there are 8 additions). Even though computers are very fast, we might wish to cut down on this number of steps. Here's how we can do it. We write our polynomial in *nested form:*

(2) F(X)=X*(X*(X*(X*(X*(X*(X*(X*9+8)+7)+6)+5)+4)+3)+2)+1

To evaluate this, start in the middle. The sequence of operations is then:

Multiply 9 by X
add 8
multiply by X

add 7
multiply by X

add 6
and so on

No values have to be stored for future use. If you follow these steps through to the end, you will find that only 16 operations are needed for each value of X.

8C-2: NESTED-EVAL For X=1 TO 100, evaluate polynomial (1) and then the polynomial written in form (2), using a program similar to program 8C-1. In each case use DEF FNA(X). If your computer prints out the amount of "central processor time" used, see if there is any difference in time taken between using (1) and using (2).

RECURSIVE EVALUATION

8C-3: RECURS-EVAL Write a program that allows the user to input the *degree* and *coefficients* of a polynomial, and a value of X. The program should then use the nested form to evaluate P(X). In the sample solution and RUN below, notice that the quantity P gets changed many times as the program loops through lines 120 and 130, with each new value of P computed in terms of its previous value. This is called *recursive evaluation.* Another example of the power of recursive evaluation is given in program 9E-1.

```
10  PRINT "WHAT IS THE DEGREE OF YOUR POLYNOMIAL (<=9)";
20  INPUT D
30  FOR I=1 TO D+1
40  PRINT "      COEFFICIENT OF X↑";D+1-I;" IS";
50  INPUT C[I]
60  NEXT I
70  PRINT "WHAT IS X (TYPE 9E9 TO STOP)";
80  INPUT X
90  IF X>8.E+09 THEN 170
100  LET P=C[1]
110  FOR I=2 TO D+1
120  LET P=P*X
130  LET P=P+C[I]
140  NEXT I
150  PRINT "P(";X;") =";P
160  GOTO 70
170  END
```

These steps correspond to the sequence of operations shown above.

```
RUN


WHAT IS THE DEGREE OF YOUR POLYNOMIAL (<=9)?8
     COEFFICIENT OF X↑ 8 IS?9
     COEFFICIENT OF X↑ 7 IS?8
     COEFFICIENT OF X↑ 6 IS?7
     COEFFICIENT OF X↑ 5 IS?6
     COEFFICIENT OF X↑ 4 IS?5
     COEFFICIENT OF X↑ 3 IS?4
     COEFFICIENT OF X↑ 2 IS?3
     COEFFICIENT OF X↑ 1 IS?2
     COEFFICIENT OF X↑ 0 IS?1
WHAT IS X (TYPE 9E9 TO STOP)?1
P( 1) = 45
WHAT IS X (TYPE 9E9 TO STOP)?2
P( 2) = 4097
WHAT IS X (TYPE 9E9 TO STOP)?-1
P(-1) = 5
WHAT IS X (TYPE 9E9 TO STOP)?9E9
```

8C-4: POLY-GRAPH Notice that in program 7D-1, Y had to be defined in two places, line 60 and line 230. If we use DEF FNA(X), one step will serve for the entire program. Compare program 7D-1 with the following program, where we also use FNA(X) for other purposes:

```
10  LET A=1
20  LET B=11
30  DEF FNA(X)=-X↑2+12*X
40  LET R1=FNA(A)
50  LET R2=FNA(A)
60  FOR X=A+1 TO B
70  IF R1 <= FNA(X) THEN 90
80  LET R1=FNA(X)
90  IF R2 >= FNA(X) THEN 110
100  LET R2=FNA(X)
110  NEXT X
120  IF 2*(R2-R1)>56 THEN 270
130  FOR I=R1 TO R2+2 STEP 2
140  PRINT TAB(7+2*(I-R1));I;
150  NEXT I
160  PRINT
170  PRINT TAB(5);"---+";
180  FOR I=1 TO (R2+2-R1)/2
190  PRINT "---+";
200  NEXT I
210  PRINT "---Y"
220  FOR X=A TO B
230  PRINT X;TAB(8+2*(FNA(X)-R1));"*"
240  NEXT X
250  PRINT " X"
260  STOP
270  PRINT "TOO WIDE--R1 =";R1;" AND R2 =";R2
280  END
```

By changing lines 10–30, you can try this program for other functions. By changing line 220, you can spread the graph in the X direction. (This distorts the graph, of course.)

Here is one sequence of experiments.

```
10  LET A=-3
20  LET B=3
30  DEF FNA(X)=X↑3
RUN


TOO WIDE--R1 =-27 AND R2 = 27

END
10  LET A=-2
20  LET B=2
RUN


       -8  -6  -4  -2   0   2   4   6   8   10
     ---+---+---+---+---+---+---+---+---+---+---Y
-2      *
-1                    *
 0                      *
 1                        *
 2                                      *
 X

END
220  FOR X=A TO B STEP .5
RUN


       -8  -6  -4  -2   0   2   4   6   8   10
     ---+---+---+---+---+---+---+---+---+---+---Y
-2      *
-1.5             *
-1                    *
-.5                    *
 0                      *
 .5                     *
 1                        *
 1.5                          *
 2                                      *
 X

END
```

Experiments of this kind will help you discover the best domain and step size for a good graph. Here's an example of a finished product:

```
10  LET A=-3
20  LET B=3
30  DEF FNA(X)=X*X*(X*X-8)+15
220  FOR X=A TO B STEP .25
```

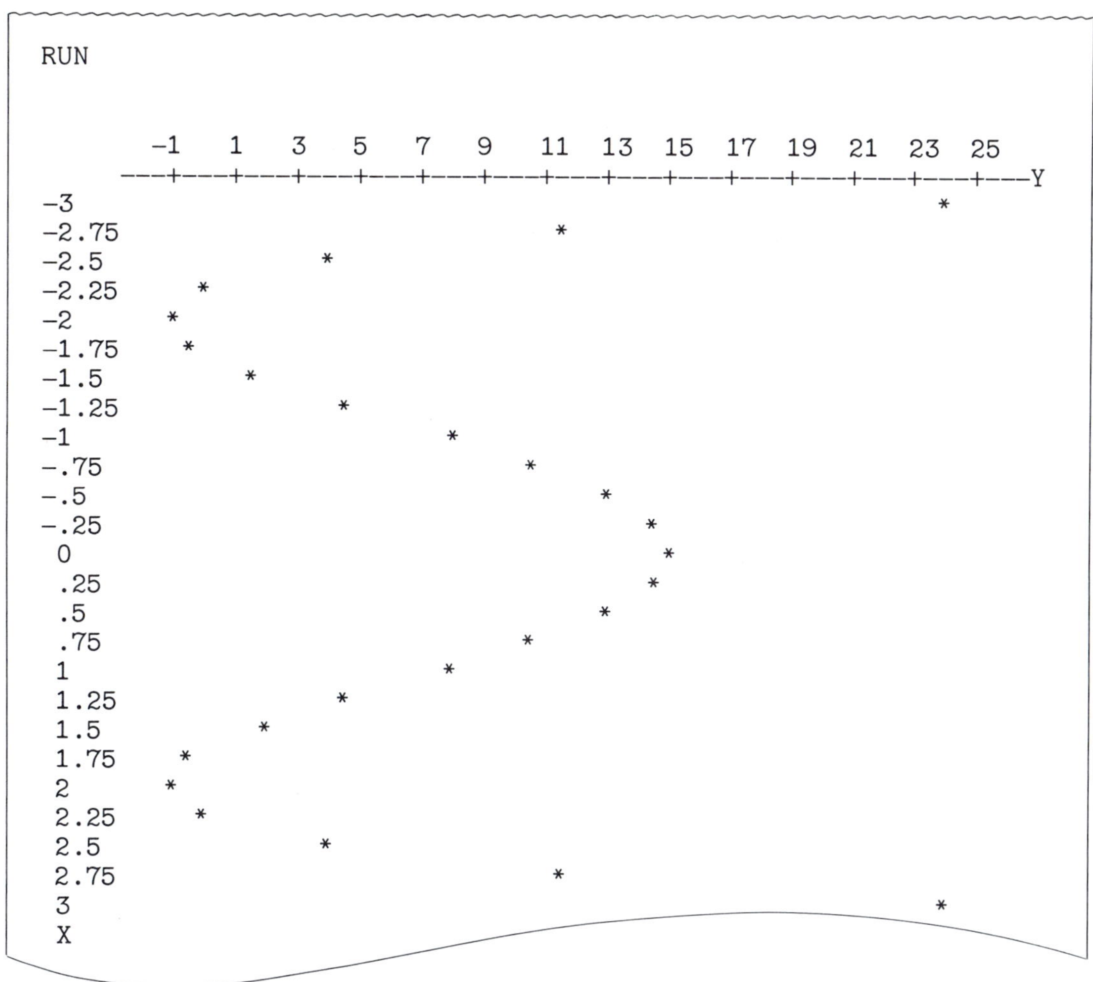

```
RUN


       -1   1   3   5   7   9  11  13  15  17  19  21  23  25
     ---+---+---+---+---+---+---+---+---+---+---+---+---+---+---Y
-3                                                        *
-2.75                            *
-2.5              *
-2.25     *
-2      *
-1.75    *
-1.5         *
-1.25              *
-1                        *
-.75                           *
-.5                                 *
-.25                                   *
 0                                      *
 .25                                   *
 .5                                 *
 .75                           *
 1                        *
 1.25              *
 1.5          *
 1.75    *
 2      *
 2.25     *
 2.5              *
 2.75                            *
 3                                                        *
 X
```

Unit 8D. Finding Roots; Search Strategies

The cab driver in our picture is using a "search strategy" to find the house at number 243. The dotted lines show how he successively narrows down the position of the exact address he wants.

The same idea provides a very powerful technique for finding the roots of an equation (the method is usually not suggested in algebra books because it would be far too tedious without a computer). For example, suppose that we are to find the real roots of the equation:

$$f(x) = 2x^3 + 6x^2 - 18x + 6 = 0$$

We first locate the roots between integers. One way is to evaluate the polynomial for several values of x and find where the graph crosses the x-axis:

```
5  PRINT " X"," F(X)"
10  DEF FNP(X)=X*(X*(X*2+6)-18)+6
20  FOR X=-5 TO 3
30  PRINT X,FNP(X)
40  NEXT X
50  END
RUN


 X                  F(X)
-5                 -4
-4                  46
-3                  60
-2                  50
-1                  28
 0                  6
 1                 -4
 2                  10
 3                  60
```

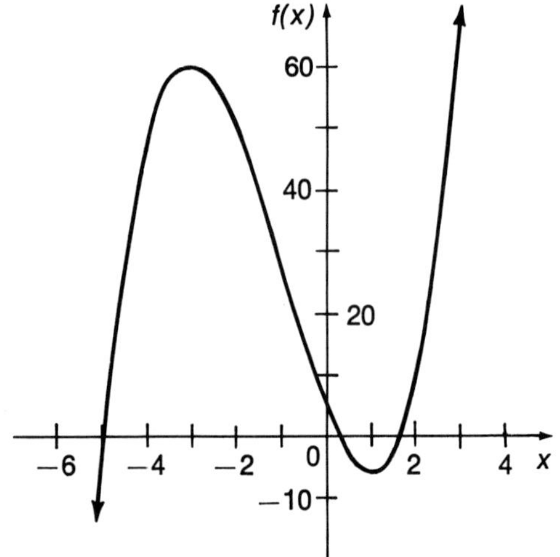

From this graph we can see that our equation has three real roots (a root is a value of X that makes $f(\mathrm{X})=0$). The best we can say at this point is that one root is in $\{-5 \leq \mathrm{X} \leq -4\}$, another is in $\{0 \leq \mathrm{X} \leq 1\}$, and the third is in $\{1 \leq \mathrm{X} \leq 2\}$.

NOTE: A *root* of the equation $f(\mathrm{X})=0$ is also called a *zero* of $f(\mathrm{X})$.

You can also get the computer to locate these roots for you.

8D-1: LOC-ROOTS Write a program that will locate the roots between integers. Here is a sample solution:

```
10  LET K=0
20  DEF FNP(X)=X*(X*(X*2+6)-18)+6
30  FOR X=-5 TO 5
40  IF FNP(X)*FNP(X+1)>0 THEN 100   <-- Tests for same sign.
50  IF FNP(X)=0 THEN 80
60  PRINT "ROOT BETWEEN";X;" AND";X+1
70  GOTO 90
80  PRINT X;" IS A ROOT."
90  LET K=K+1
100  NEXT X
110  IF K>0 THEN 130
120  PRINT "NO ROOTS LOCATED"
130  END
RUN


ROOT BETWEEN-5 AND-4
ROOT BETWEEN 0 AND 1
ROOT BETWEEN 1 AND 2
```

Our problem now is to get a more accurate value for each of these roots. We'll do this by conducting what is called a "binary search" in each interval. This means that we'll cut each interval in half, and then use the computer to find out which half contains the root. Then we'll cut *that* interval in half (now we're down to an interval of only $\frac{1}{4}$ (or 2↑−2) the original size). If we repeat the process say 20 times, we'll have the root within 2↑−20, which means with an accuracy of about ±.000001 units.

8D-2: BINARY-SEARCH Having located the roots between integers, let's draw a picture of what a binary search would look like in the interval $\{1 \leq X \leq 2\}$. The colored segments represent the first 7 successive intervals.

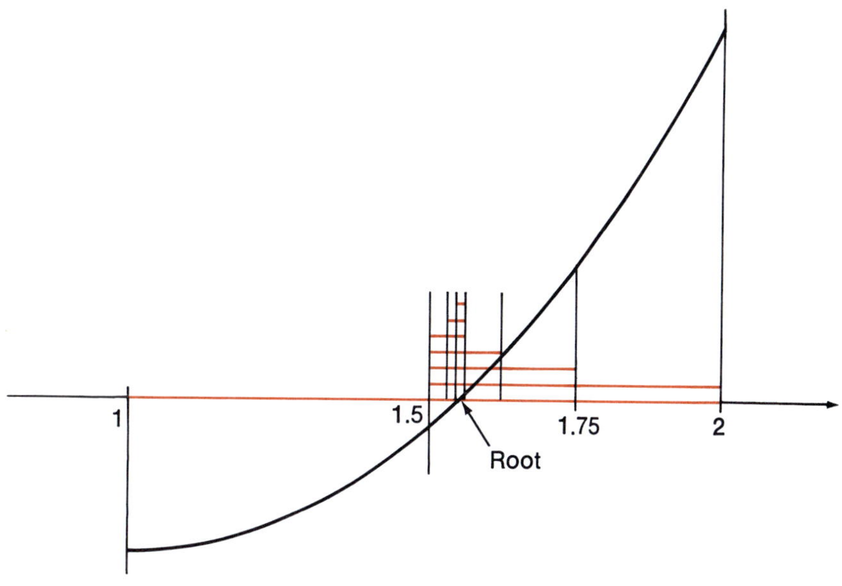

The rule for selecting each new "half-interval" is to choose it so that $f(X)$ is negative at one end and positive at the other. This ensures that the graph will cross the X-axis within our new interval—that we are "homing" in on a root. At each stage, call the value of X at the left end point A and the value of X at the right end point B. Then the value of X at the midpoint is:

$$M = (A + B)/2$$

Here's a program to do this, together with a RUN:

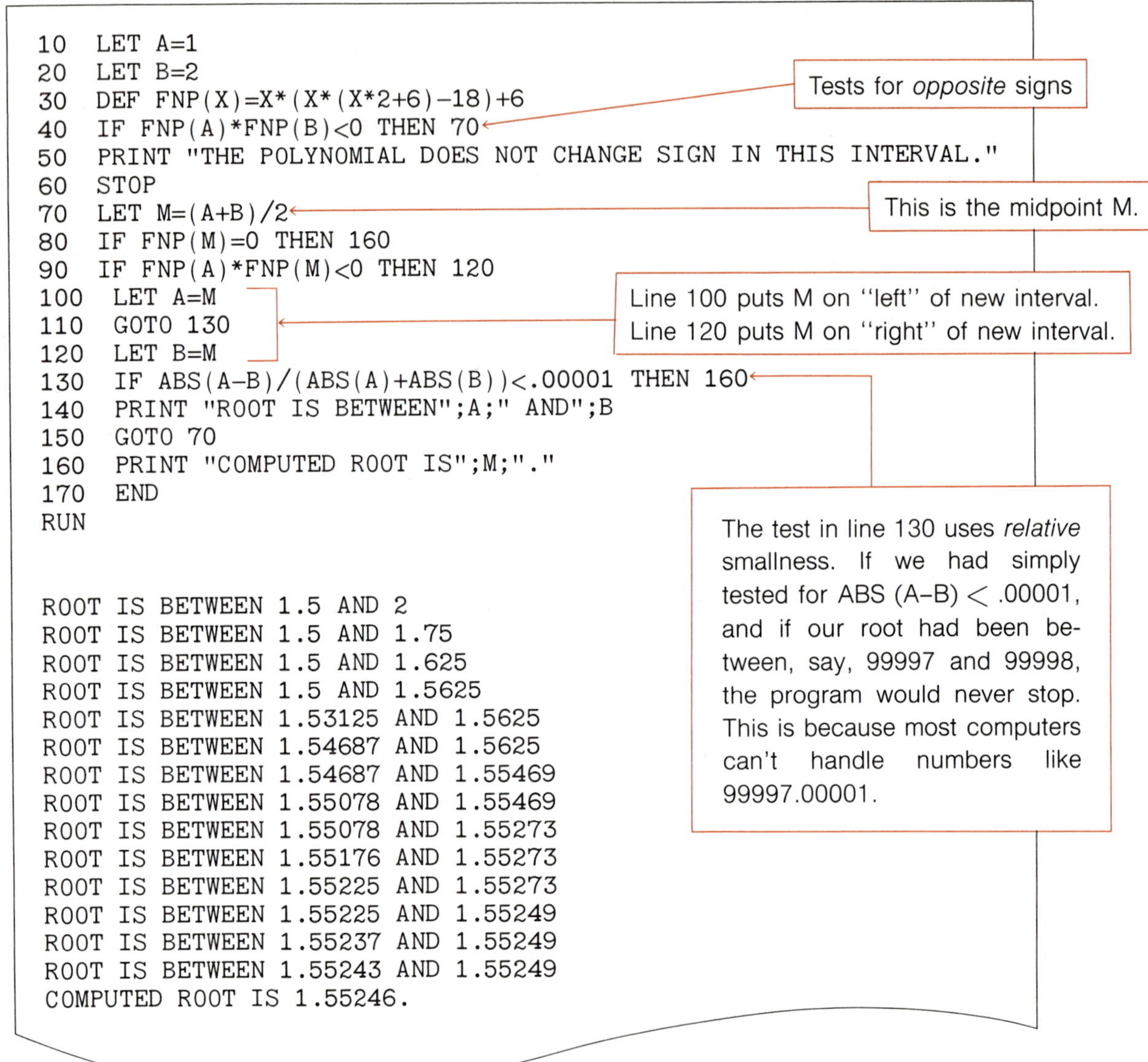

```
10 LET A=1
20 LET B=2
30 DEF FNP(X)=X*(X*(X*2+6)-18)+6
40 IF FNP(A)*FNP(B)<0 THEN 70
50 PRINT "THE POLYNOMIAL DOES NOT CHANGE SIGN IN THIS INTERVAL."
60 STOP
70 LET M=(A+B)/2
80 IF FNP(M)=0 THEN 160
90 IF FNP(A)*FNP(M)<0 THEN 120
100 LET A=M
110 GOTO 130
120 LET B=M
130 IF ABS(A-B)/(ABS(A)+ABS(B))<.00001 THEN 160
140 PRINT "ROOT IS BETWEEN";A;" AND";B
150 GOTO 70
160 PRINT "COMPUTED ROOT IS";M;"."
170 END
RUN


ROOT IS BETWEEN 1.5 AND 2
ROOT IS BETWEEN 1.5 AND 1.75
ROOT IS BETWEEN 1.5 AND 1.625
ROOT IS BETWEEN 1.5 AND 1.5625
ROOT IS BETWEEN 1.53125 AND 1.5625
ROOT IS BETWEEN 1.54687 AND 1.5625
ROOT IS BETWEEN 1.54687 AND 1.55469
ROOT IS BETWEEN 1.55078 AND 1.55469
ROOT IS BETWEEN 1.55078 AND 1.55273
ROOT IS BETWEEN 1.55176 AND 1.55273
ROOT IS BETWEEN 1.55225 AND 1.55273
ROOT IS BETWEEN 1.55225 AND 1.55249
ROOT IS BETWEEN 1.55237 AND 1.55249
ROOT IS BETWEEN 1.55243 AND 1.55249
COMPUTED ROOT IS 1.55246.
```


PROBLEMS: Find the real roots of:

1. $x^3 - 2 = 0$
2. $x^4 - 2.4x^3 + 1.03x^2 + .6x - .32 = 0$
3. $x^3 + 2x^2 + 10x - 20 = 0$

Check your answers by substitution (use 8C-3).

Unit 8E. Polynomial Fortune Tellers

In Section 5-E we used a linear fortune teller with the linear equation in the form

$$y = mx + b.$$

In Section 6-D we found that the equation of a line through points $P_1:(x_1, y_1)$ and $P_2:(x_2, y_2)$ could be written in the form:

$$y = \left(\frac{y_2 - y_1}{x_2 - x_1}\right)x + \frac{x_2y_1 - x_1y_2}{x_2 - x_1}$$

This can be rearranged into the form:

$$y = \frac{(x - x_2)}{(x_1 - x_2)}y_1 + \frac{(x - x_1)}{(x_2 - x_1)}y_2$$

This latter equation is called the *Lagrangian form* of the first-degree polynomial equation through P_1 and P_2. This is named after Joseph Louis Lagrange, an 18th-century French mathematician.

8E-1: LAGRANGE-1 Rewrite program 5E-3, using the Lagrangian form of a linear equation, as given above.

However, not all situations in which we want to interpolate or extrapolate involve linear functions. Sometimes functions defined by higher-degree polynomial equations must be used. Let's see how this works for second-degree polynomials.

Two points determine a *straight line,* which is described by a linear equation of the form

$$y = a_1x + a_2.$$

Three points determine a *parabola,* which is described by a second-degree equation of the form

$$y = a_1x^2 + a_2x + a_3.$$

Here is the Lagrangian form of the second-degree equation which is determined when three points, (x_1, y_1), (x_2, y_2), (x_3, y_3), are given:

$$y = \frac{(x - x_2)(x - x_3)}{(x_1 - x_2)(x_1 - x_3)}y_1 + \frac{(x - x_1)(x - x_3)}{(x_2 - x_1)(x_2 - x_3)}y_2 + \frac{(x - x_1)(x - x_2)}{(x_3 - x_1)(x_3 - x_2)}y_3$$

8E-2: LAGRANGE-2 Write a program that uses data from *three* points to find Y at a fourth point. A sample program followed by a RUN is shown on page 144.

```
10  PRINT "INPUT FIRST PAIR OF COORDINATES: X1, Y1";
20  INPUT X1,Y1
30  PRINT "INPUT SECOND PAIR OF COORDINATES: X2, Y2";
40  INPUT X2,Y2
50  PRINT "INPUT THIRD PAIR OF COORDINATES: X3, Y3";
60  INPUT X3,Y3
70  PRINT "INPUT A VALUE OF X FOR WHICH YOU WANT Y TO BE ESTIMATED";
80  INPUT X
90  LET P=Y1*((X-X2)*(X-X3))/((X1-X2)*(X1-X3))
100  LET P=P+Y2*((X-X1)*(X-X3))/((X2-X1)*(X2-X3))
110  LET P=P+Y3*((X-X1)*(X-X2))/((X3-X1)*(X3-X2))
120  PRINT "FOR X =";X;", Y =";P
130  END
RUN


INPUT FIRST PAIR OF COORDINATES: X1, Y1?20,50
INPUT SECOND PAIR OF COORDINATES: X2, Y2?30,40
INPUT THIRD PAIR OF COORDINATES: X3, Y3?40,60
INPUT A VALUE OF X FOR WHICH YOU WANT Y TO BE ESTIMATED?45
FOR X = 45, Y = 81.25
```

8E-3: CAR The table below shows the time T (in seconds) required for a car to accelerate from 0 to various velocities V (in kilometers per hour). Notice that the elapsed-time figures are missing in the table for V = 10, V = 20, V = 40, and V = 50. (T is *actually* missing for *all* values of V *except* 0, 30, and 60. . . . Think about it.)

V (km/h)	T (seconds)
0	0
10	—
20	—
30	2.7
40	—
50	—
60	5.1

For purposes of comparison with other cars, it might be important, for example, to know the times for accelerating from zero to thirty and from zero to sixty.

We might consider using linear interpolation or extrapolation. However, we notice that T is not a linear function of V, since doubling V (30 to 60) doesn't double T (2.7 to 5.1). So it will be better to use program 8E-1 (second-degree interpolation). Do this to obtain the values of V that are missing in the table.

SECTION 9 POLYNOMIAL FUNCTIONS AND COMPLEX NUMBERS

Checklist of Computing Skills

Previously explained	PRINT, END, LET, INPUT, IF . . . THEN, STOP, GOTO, FOR, NEXT, SQR, ABS, PRINT TAB Subscripted variables—single subscripts, RND GOSUB, RETURN
Requires extra reading	String variables (Program 9B-3)

Checklist of Algebraic Skills

1. $(6x^3 + 2x^2 - 3x + 4) + (2x^2 + 7x - 3) =$ __?__
2. $(6x^3 + 2x^2 - 3x + 4)(2x^2 + 7x - 3) =$ __?__
3. The two roots of $x^2 + x + 1 = 0$ are __?__ and __?__.
4. $(x^3 - 9x^2 + 26x - 24)/(x - 2) =$ __?__; remainder, __?__
5. If you found in Exercise 4 that 2 is a root of

$$x^3 - 9x^2 + 26x - 24 = 0,$$

use the quadratic formula to find the other two roots. Then:

$$x^3 - 9x^2 + 26x - 24 = (x - 2)(x - \_\_?\_\_)(x - \_\_?\_\_)$$

Check your answers with those printed at the bottom of the page.

Unit 9A. EXAM-CRAM, Part V

9A-1: POLYMULT Write a program to provide practice in multiplying polynomials. Below is a sample RUN. Get the program to run and have several students try it. Ask for their comments and criticisms, and use these to improve the program.

```
RUN



THIS SECTION TESTS YOUR ABILITY TO MULTIPLY TWO POLYNOMIALS.
YOU WILL BE ASKED TO GIVE THE COEFFICIENT OF EACH TERM IN
THE PRODUCT.  USE PENCIL AND PAPER AS NECESSARY.

FIND THE PRODUCT OF:
 7X↑ 5 + 7X↑ 4 + ( 3)X↑ 3
   AND    7X↑ 3 + 7X↑ 2
THE COEFFICIENT OF X↑ 8 IS:?49
THAT'S RIGHT
THE COEFFICIENT OF X↑ 7 IS:?28
NO; THE COEFFICIENT IS: 98
THE COEFFICIENT OF X↑ 6 IS:?70
THAT'S RIGHT
THE COEFFICIENT OF X↑ 5 IS:?10
NO; THE COEFFICIENT IS: 21
```

Answers: (1) $6x^3 + 4x^2 + 4x + 1$ (2) $12x^5 + 46x^4 - 10x^3 - 19x^2 + 37x - 12$ (3) $-\frac{1}{2} + \frac{1}{2}\sqrt{3}\,i$, $-\frac{1}{2} - \frac{1}{2}\sqrt{3}\,i$ (4) $x^2 - 7x + 12$, 0 (5) 3, 4

```
WOULD YOU LIKE TO DO ANOTHER (1=YES, 0=NO)?1

FIND THE PRODUCT OF:
 8X↑ 3 + 3X↑ 2 + (-2)X↑ 1
  AND   5X↑ 1 + 4X↑ 0
THE COEFFICIENT OF X↑ 4 IS:?40
THAT'S RIGHT
THE COEFFICIENT OF X↑ 3 IS:?47
THAT'S RIGHT
THE COEFFICIENT OF X↑ 2 IS:?22
NO; THE COEFFICIENT IS: 2
THE COEFFICIENT OF X↑ 1 IS:?8
NO; THE COEFFICIENT IS: -8

WOULD YOU LIKE TO DO ANOTHER (1=YES, 0=NO)?0
```

9A-2: POLYOP Write a program to practice adding and subtracting polynomials.

9A-3: COMPLEX Write a program that gives practice in adding and multiplying complex numbers. Here is a sample RUN.

```
RUN


THIS SECTION TESTS YOUR ABILITY TO ADD OR MULTIPLY
TWO COMPLEX NUMBERS.  INPUT YOUR ANSWER, 'M+NI,' AS
M,M.  USE PENCIL AND PAPER AS NECESSARY.

WHAT IS THE PRODUCT OF -8+( 1)I AND 7+( 3)I?-59,-17
THAT'S RIGHT

WOULD YOU LIKE TO DO ANOTHER (1=YES, 0=NO)?1

WHAT IS THE PRODUCT OF -5+(-3)I AND 0+(-6)I?13,30
NO; THE PRODUCT IS -18+( 30)I

WOULD YOU LIKE TO DO ANOTHER (1=YES, 0=NO)?1

WHAT IS THE SUM OF -7+(-5)I AND 1+( 7)I?-6,2
THAT'S RIGHT

WOULD YOU LIKE TO DO ANOTHER (1=YES, 0=NO)?0
```

Unit 9B. EXAM-CRAM, Part VI

9B-1: COMBO-2 Combine programs 7A-2, 7A-5, 8A-1, and 9A-1 (or any other combination you wish) into one long program. By dropping the END statements from 7A-2, 7A-5, and 8A-1, a "NO" to the "do another" question should cause each program to continue on to the next higher line number (in the next program). Some lines will have to be reworded slightly. If necessary, renumber the lines in later programs so that there will be no overlaps. Perhaps your system has a command that will do this for you.

9B-2: SMART-COMPLEX Modify the program 9A-3 so that when a wrong answer is given, the program prints separate explanations of how the real and imaginary parts should be calculated.

9B-3: STRING-COMPLEX If your computer can handle *string variables,* rewrite program 9A-3 so that it will accept INPUT in the form 3+5I, and so on.

9B-4: QUADSOLVE-3 Modify program 7C-1 to print out complex roots when $B^2 - 4AC$ is negative. For example, for A = 1, B = 2, and C = 3, your program should print:

```
THE TWO ROOTS ARE:
R1 = -.5+1.41421I
R2 = -.5-1.41421I
```

Unit 9C. Synthetic Division

You may have learned in your algebra textbook how to divide a polynomial

$$P(x) = 4x^3 - 2x^2 - 2$$

by a binomial

$$x - 2$$

using *synthetic division* as shown at left.

The sequence of operations to get these numbers is as follows:

2	4	−2	0	−2
		8	12	24
	4	6	12	22

This means that the quotient is $4x^2 + 6x + 12$ and the remainder is 22.

Copy down the first coefficient, 4
multiply 4 by 2
add −2
multiply by 2
add 0
multiply by 2
add −2

Compare these operations with those on page 136. They are exactly the same!

In other words, dividing $P(x)$ by $x - 2$ gives a remainder equal to what we get by evaluating $P(2)$. This is called the Remainder Theorem.

9C-1: SYNTHDIV Write a program that uses synthetic division to find the quotient and remainder when a polynomial of any degree is divided by a linear (first degree) polynomial of the form X − B.

A sample solution, with a RUN is shown on page 148.

```
10  PRINT "WHAT IS THE DEGREE OF YOUR POLYNOMIAL (<=9)";
20  INPUT D
30  PRINT "INPUT THE COEFFICIENTS OF DESCENDING POWERS OF X:"
40  FOR I=1 TO D+1
50  INPUT C[I]
60  NEXT I
70  PRINT
80  PRINT "WHAT IS THE CONSTANT B IN THE DIVISOR (X-B)";
90  INPUT B
100  LET A[1]=C[1]
110  FOR I=2 TO D+1
120  LET A[I]=A[I-1]*B+C[I]
130  NEXT I
140  PRINT
150  PRINT "QUOTIENT IS: ";
160  PRINT A[1];"X↑";D-1;
170  FOR I=2 TO D
180  IF A[I]>0 THEN 210
190  PRINT " -";ABS(A[I]);"X↑";D-I;
200  GOTO 220
210  PRINT " +";A[I];"X↑";D-I;
220  NEXT I
230  PRINT
240  PRINT "REMAINDER, P(B), IS: ";A[D+1]
250  PRINT
260  PRINT "WOULD YOU LIKE TO TRY ANOTHER VALUE (1=YES, 0=NO)";
270  INPUT K
280  IF K=1 THEN 70
290  END
RUN


WHAT IS THE DEGREE OF YOUR POLYNOMIAL (<=9)?3
INPUT THE COEFFICIENTS OF DESCENDING POWERS OF X:
?4
?-2
?0
?-2

WHAT IS THE CONSTANT B IN THE DIVISOR (X-B)?2

QUOTIENT IS:   4X↑ 2 + 6X↑ 1 + 12X↑ 0
REMAINDER, P(B), IS:  22

WOULD YOU LIKE TO TRY ANOTHER VALUE (1=YES, 0=NO)?1

WHAT IS THE CONSTANT B IN THE DIVISOR (X-B)?1

QUOTIENT IS:   4X↑ 2 + 2X↑ 1 + 2X↑ 0
REMAINDER, P(B), IS:  0

WOULD YOU LIKE TO TRY ANOTHER VALUE (1=YES, 0=NO)?0
```

This line corresponds to lines 120 and 130 in program 8C-3. Here all the intermediate values are stored in A(I).

You know from the Factor Theorem that you learned in algebra that when the remainder is zero, X − B is a factor of P(X) and B is a root of the equation P(X) = 0. Thus, in the second part of the RUN above, we see that 1 is a root of $4x^3 - 2x^2 - 2 = 0$.

Unit 9D. The Ultramatic Root-Finder

In this unit we're going to bring together some of the techniques developed in programs 8C-3 (polynomial evaluation), 8D-2 (finding roots by binary search), 9B-4 (the quadratic equation), and 9C-1 (synthetic division). Our goal is to produce a flexible, professional root-finding program.

Program 9C-1 was used to find that 1 is a root of the equation

$$4x^3 - 2x^2 - 2 = 0$$

and that when $4x^3 - 2x^2 - 2$ was divided by $x - 1$, the quotient was

$$4x^2 + 2x + 2.$$

You know from your work in algebra that any other root of the original equation must be a root of the *depressed equation*

$$4x^2 + 2x + 2 = 0.$$

This idea can be used in a root-finding program.

9D-1: MULTI-ROOT Write a program that allows the user to input the degree and coefficients of a polynomial equation. The program should then search for roots in an interval specified by the user, depressing the equation if roots are found. Here is a sample solution that combines steps from programs 8C-3, 8D-2, and 9C-1. This program stops if a quadratic equation is reached in the depressing process.

From 9C-1 (lines 30–80)

Locates real roots in a specified interval. (lines 100–260)

```
10  PRINT "SEARCH FOR SOME ROOTS OF A POLYNOMIAL EQUATION"
20  PRINT
30  PRINT "WHAT IS THE DEGREE OF THE POLYNOMIAL (3<=D<=9)";
40  INPUT D
50  PRINT "INPUT THE COEFFICIENTS OF DESCENDING POWERS OF X:"
60  FOR I=1 TO D+1
70  INPUT C[I]
80  NEXT I
90  PRINT
100  PRINT "SPECIFY INTERVAL TO BE SEARCHED:"
110  PRINT "WHAT IS THE LOWER BOUND";
120  INPUT L
130  PRINT "WHAT IS THE UPPER BOUND";
140  INPUT U
150  LET X=L
160  GOSUB 600
170  IF P=0 THEN 400
180  LET K1=P
190  FOR K=L+1 TO U
200  LET X=K
210  GOSUB 600
220  IF K1*P<0 THEN 280
230  IF P=0 THEN 400
240  LET K1=P
250  NEXT K
260  PRINT "NO ROOTS IN THE INTERVAL FROM ";L;" TO ";U
270  STOP
```

From 8D-2

```
280 LET A=K-1
290 LET B=K
300 PRINT "ROOT BETWEEN ";A;" AND ";B
310 LET X=(A+B)/2
320 GOSUB 600
330 IF P=0 THEN 400
340 IF K1*P<0 THEN 370
350 LET A=X
360 GOTO 380
370 LET B=X
380 IF ABS(A-B)/(ABS(A)+ABS(B))<.00001 THEN 400
390 GOTO 310
400 PRINT X;" IS A ROOT."
```

From 9C-1

```
410 LET A[1]=C[1]
420 FOR I=2 TO D+1
430 LET A[I]=A[I-1]*X+C[I]
440 NEXT I
```

Sets up depressed equation.

```
450 LET D=D-1
460 PRINT
470 PRINT "COEFFICIENTS OF DEPRESSED EQUATION ARE:"
480 FOR I=1 TO D+1
490 LET C[I]=A[I]
500 PRINT C[I];" ";
510 NEXT I
520 PRINT
530 IF D=2 THEN 580
```

Sets up reduced interval.

```
540 LET L=K
550 IF L=U THEN 590
560 PRINT
570 GOTO 150
580 PRINT "SOLVE QUADRATIC EQUATION."
590 STOP
```

From 8C-3

```
600 LET P=C[1]
610 FOR I=2 TO D+1
620 LET P=X*P+C[I]
630 NEXT I
640 RETURN
650 END
RUN

SEARCH FOR SOME ROOTS OF A POLYNOMIAL EQUATION

WHAT IS THE DEGREE OF THE POLYNOMIAL (3<=D<=9)?3
INPUT THE COEFFICIENTS OF DESCENDING POWERS OF X:
?4
?-2
?0
?-2

SPECIFY INTERVAL TO BE SEARCHED:
WHAT IS THE LOWER BOUND?-2
WHAT IS THE UPPER BOUND?2
 1 IS A ROOT.

COEFFICIENTS OF DEPRESSED EQUATION ARE:
 4   2   2
SOLVE QUADRATIC EQUATION.

END
```

```
RUN


SEARCH FOR SOME ROOTS OF A POLYNOMIAL EQUATION

WHAT IS THE DEGREE OF THE POLYNOMIAL (3<=D<=9)?4
INPUT THE COEFFICIENTS OF DESCENDING POWERS OF X:
?1
?-7
?9
?7
?-10

SPECIFY INTERVAL TO BE SEARCHED:
WHAT IS THE LOWER BOUND?-10
WHAT IS THE UPPER BOUND?10
-1 IS A ROOT.

COEFFICIENTS OF DEPRESSED EQUATION ARE:
 1  -8   17  -10

1 IS A ROOT.

COEFFICIENTS OF DEPRESSED EQUATION ARE:
 1  -7   10
SOLVE QUADRATIC EQUATION.

END
RUN

SEARCH FOR SOME ROOTS OF A POLYNOMIAL EQUATION

WHAT IS THE DEGREE OF THE POLYNOMIAL (3<=D<=9)?5
INPUT THE COEFFICIENTS OF DESCENDING POWERS OF X:
?1
?0
?2
?0
?0
?-1

SPECIFY INTERVAL TO BE SEARCHED:
WHAT IS THE LOWER BOUND?-5
WHAT IS THE UPPER BOUND?5
ROOT BETWEEN 0 AND 1
 .733162 IS A ROOT.

COEFFICIENTS OF DEPRESSED EQUATION ARE:
 1   .733162   2.53753   1.86042   1.36399

NO ROOTS IN THE INTERVAL FROM 1 TO 5
```

9D-2: ULTRA-ROOT Add steps from program 9B-4 to solve any remaining quadratic equation in program 9D-1.

Notice that this program can be made to locate real roots and the two complex roots if there are only two.

PROBLEMS: Find all the roots of:

1. $x^3 - 4x^2 + 5x - 2 = 0$
2. $x^4 - 6x^3 + 13x^2 - 12x + 4 = 0$
3. $x^5 - 3x^4 - 23x^3 + 41x^2 + 94x - 120 = 0$
4. Problem 2, page 142.
5. Explain why you can always find all the roots of any cubic equation with this program.

Unit 9E. The Punctilious Polynomial Predictor

Unit 8E showed how we can interpolate or extrapolate from data at two points (linear interpolation), or from data at three points (second-degree interpolation). Let's now look at the most general case—data at n points. We'll first extend the idea of Lagrangian form of a polynomial equation.

Recall that the Lagrangian polynomial equation of degree 1 is:

$$(1)\quad y = \frac{(x - x_2)}{(x_1 - x_2)}y_1 + \frac{(x - x_1)}{(x_2 - x_1)}y_2$$

and the Lagrangian polynomial equation of degree 2 is:

$$(2)\quad y = \frac{(x - x_2)(x - x_3)}{(x_1 - x_2)(x_1 - x_3)}y_1 + \frac{(x - x_1)(x - x_3)}{(x_2 - x_1)(x_2 - x_3)}y_2 + \frac{(x - x_1)(x - x_2)}{(x_3 - x_1)(x_3 - x_2)}y_3$$

The Langrangian polynomial equation of degree $n - 1$ is:

$$(3)\quad y = \frac{(x - x_2)(x - x_3)\cdots(x - x_n)}{(x_1 - x_2)(x_1 - x_3)\cdots(x_1 - x_n)}y_1 + \frac{(x - x_1)(x - x_3)\cdots(x - x_n)}{(x_2 - x_1)(x_2 - x_3)\cdots(x_2 - x_n)}y_2$$
$$+ \cdots + \frac{(x - x_1)(x - x_2)\cdots(x - x_{n-1})}{(x_n - x_1)(x_n - x_2)\cdots(x_n - x_{n-1})}y_n$$

Let's look more closely at one of these terms:

$$\frac{\overbrace{(x - x_1)(x - x_2)\cdots(x - x_{i-1})(x - x_{i+1})\cdots(x - x_n)}^{n\,-\,1\text{ factors}}}{\underbrace{(x_i - x_1)(x_i - x_2)\cdots(x_i - x_{i-1})(x_i - x_{i+1})\cdots(x_i - x_n)}_{n\,-\,1\text{ factors}}}y_i$$

Notice that $(x - x_i)$ is missing from the product in the numerator and $(x_i - x_i)$ is missing from the product in the denominator.

This term is equal to

$$0 \cdot y_i \qquad \text{for} \qquad x \neq x_i$$

and is equal to

$$1 \cdot y_i \qquad \text{for} \qquad x = x_i.$$

If the factors in the numerator are multiplied, the product is a polynomial in x of degree $n - 1$. The denominator is a constant and y_i is a constant, and since a constant times a polynomial is a polynomial of the same degree, the entire term is a polynomial of degree $n - 1$. Finally, the entire right-hand side of (3) is the sum of n polynomials of degree $n - 1$. But the sum of polynomials of degree $n - 1$ is a polynomial of degree less than or equal to $n - 1$. Hence:

1. Equation (3) is a polynomial equation of degree less than or equal to $n - 1$.

2. The points $(x_1, y_1), \ldots, (x_n, y_n)$ all satisfy equation (3).

9E-1 LAGRANGE-N Suppose that we have data from five sightings of a moving object—say a UFO—as shown in our picture.

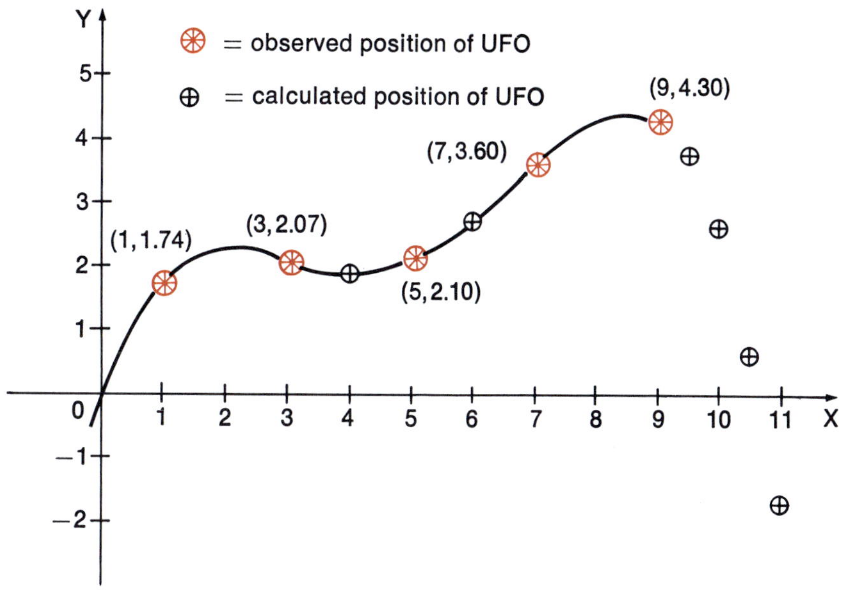

Our problem is: Given the observed values of Y at X = 1, 3, 5, 7, and 9 as shown, find Y at other points. For example, find Y(4), Y(6), Y(9.5), Y(10), Y(10.5), and Y(11). One way to do this is to find the 4th-degree Lagrangian polynomial equation through the given five points, and then calculate Y as the value of this polynomial at the points marked.

A RUN and a listing of a sample solution follow. Our program uses recursive evaluation to find the coefficients (line 190), and the value of the polynomial (line 210).

```
RUN


HOW MANY DATA POINTS DO YOU HAVE?5
NOW INPUT 5 (X,Y) PAIRS (DATA POINTS)
?1,1.74
?3,2.07
?5,2.10
?7,3.60
?9,4.30
FOR WHAT VALUE OF X DO YOU WANT Y TO BE ESTIMATED (TYPE 9E9 TO STOP):

X=?4
FOR X = 4, Y = 1.91719
```

```
X =?6
FOR X = 6, Y = 2.71344

X =?9.5
FOR X = 9.5, Y = 3.69932

X =?10
FOR X = 10, Y = 2.53594

X = ?10.5
FOR X = 10.5, Y = .656034

X = ?11
FOR X = 11, Y =-2.11

X =?9E9
```

```
10  DIM X[20],Y[20],B[20]
20  PRINT "HOW MANY DATA POINTS DO YOU HAVE";
30  INPUT N
40  PRINT "NOW INPUT";N;" (X,Y) PAIRS (DATA POINTS)"
50  FOR I=1 TO N
60  INPUT X[I],Y[I]
70  NEXT I
80  PRINT "FOR WHAT VALUE OF X DO YOU WANT Y TO BE ESTIMATED";
90  PRINT " (TYPE 9E9 TO STOP):"
100  PRINT
110  PRINT "X =";
120  INPUT X
130  IF X>8.E+09 THEN 250
140  LET P=0
150  FOR K=1 TO N
160  LET B[K]=1
170  FOR J=1 TO N
180  IF J=K THEN 200
190  LET B[K]=B[K]*(X-X[J])/(X[K]-X[J])
200  NEXT J
210  LET P=P+B[K]*Y[K]
220  NEXT K
230  PRINT "FOR X =";X;", Y =";P
240  GOTO 100
250  END
```

9E-2: GRAPH-PREDICT Extend program 9E-1 to graph the (N − 1)-degree polynomial equation, using different symbols for data and calculated values.

SECTION 10 COMPUTER-GENERATED ANIMATION

Checklist of Computing Skills

Previously explained	PRINT, END, LET, INPUT, IF . . . THEN, STOP, GOTO, FOR, NEXT, SQR, INT, PRINT TAB, COMPUTED GOTO, RND

Checklist of Algebraic Skills

1. 3 is __?__% of 5.
2. 5% of 200 is __?__.
3. 5 is 10% of __?__.
4. If a pump can fill a tank in 3 hours, what part of the contents has been put into the tank in 1 hour?
5. If $x = 2t$ and $y = 5t^2$, solve for t in terms of x in the first equation and substitute in the second equation to obtain an equation in x and y.

Check your answers with those printed at the bottom of the page.

Unit 10A. EXAM-CRAM, Part VII

10A-1: PERCENT Write a program to drill a student in percentage problems. Select the three types at random.

Here is a sample RUN:

```
RUN


THIS SECTION TESTS YOUR ABILITY TO DO PERCENTAGE
PROBLEMS.  USE PENCIL AND PAPER AS NECESSARY.

 22.6667 IS 66.6667% OF WHAT NUMBER?34
CORRECT

WOULD YOU LIKE TO DO ANOTHER (1=YES, 0=NO)?1

WHAT PERCENT OF 22 IS 55?250
CORRECT

WOULD YOU LIKE TO DO ANOTHER (1=YES, 0=NO)?1

 45 IS 166.667% OF WHAT NUMBER?18
NO.  B = 27

WOULD YOU LIKE TO DO ANOTHER (1=YES, 0=NO)?1

 125% OF 54 IS WHAT NUMBER?67.5
CORRECT

WOULD YOU LIKE TO DO ANOTHER (1=YES, 0=NO)?1
```

Answers: (1) 60% (2) 10 (3) 50 (4) $\frac{1}{3}$ (5) $y = \frac{5}{4}x^2$

```
 66 IS 200% OF WHAT NUMBER?33
CORRECT

WOULD YOU LIKE TO DO ANOTHER (1=YES, 0=NO)?1

 25% OF 33 IS WHAT NUMBER?8.25
CORRECT

WOULD YOU LIKE TO DO ANOTHER (1=YES, 0=NO)?1

WHAT PERCENT OF 59 IS 44.25?82
NO.  R = 75

WOULD YOU LIKE TO DO ANOTHER (1=YES, 0=NO)?1

 28 IS 133.333% OF WHAT NUMBER?21
CORRECT

WOULD YOU LIKE TO DO ANOTHER (1=YES, 0=NO)?1

 25% OF 45 IS WHAT NUMBER?11.5
NO.  P = 11.25

WOULD YOU LIKE TO DO ANOTHER (1=YES, 0=NO)?1

WHAT PERCENT OF 21 IS 5.25?25
CORRECT

WOULD YOU LIKE TO DO ANOTHER (1=YES, 0=NO)?0
```

One way of beginning such a program is:

```
PRINT
PRINT "THIS SECTION TESTS YOUR ABILITY TO DO PERCENTAGE"
PRINT "PROBLEMS.  USE PENCIL AND PAPER AS NECESSARY."
PRINT
LET K=INT(3*RND(1)+1)
LET R1=INT(5*RND(1)+1)
LET R2=INT(4*RND(1)+2)
LET R=(R1/R2)*100
LET B=INT(50*RND(1)+10)
LET P=(R1/R2)*B
GOTO K OF ..., ..., ...,
```

10A-2: PUMP-PROB Write a program to give a student practice in solving word problems of the following type:

> If pump number 1 fills a tank in 3 hours, and if pump number 2 fills the same tank in 4 hours, how long will it take to fill that tank with both pumps working together?

The underlined numbers are to be randomly generated.

Unit 10B. EXAM-CRAM, Part VIII

10B-1: COMBO-3 Combine several "EXAM-CRAM" programs into one long program and insert lines at the beginning to allow the student to choose the section he wishes to use.

For example, it might begin like this:

```
1  RANDOMIZE (Use only if necessary.)
2  PRINT "EXAM-CRAM -- THIS PROGRAM HAS SEVERAL SECTIONS:"
3  PRINT "SECTION 1:  MULTIPLYING BINOMIALS"
4  PRINT "SECTION 2:  FACTORING TRINOMIALS"
.
.
.
28  PRINT "WHICH SECTION DO YOU WISH (TYPE NUMBER OF SECTION)";
29  INPUT K
30  GOTO K OF 40,200,...
.
.
.
```

A number of lines in the separate programs will have to be changed.

WHY COMPUTER MOVIES?

To find one answer to this question, let's ask another. Can a person learn to perform a complicated and dangerous task (say landing a helicopter on a ship at sea) without harm to life and limb? Yes, provided the person practices in a simulator, a device that *imitates* the performance of the real thing. To be useful, the simulator must give the student visual cues—in our example, moving pictures of the landing pad as it would be seen from the window of a real helicopter. But movies are made up of thousands of "snapshots" called *frames,* each one a little different from the last. Producing such pictures artificially is a task for which computers are well suited.*

In the following units we'll get some idea of the principles used in generating movies with computers by writing a program that produces successive pictures of a bouncing ball. The twelve pictures shown on page 158 were produced by such a program. We'll organize our study of how to make such a movie by examining:

(1) the mathematics needed

(2) a computer program that produces the snapshots

(3) methods for producing an animation (movie) from these snapshots

*Because we are using a slow-speed terminal for output, our programs are not suitable for use in a "real time" simulator, which would require a high-speed terminal that produces output on a TV-like screen. However, the principles involved are the same.

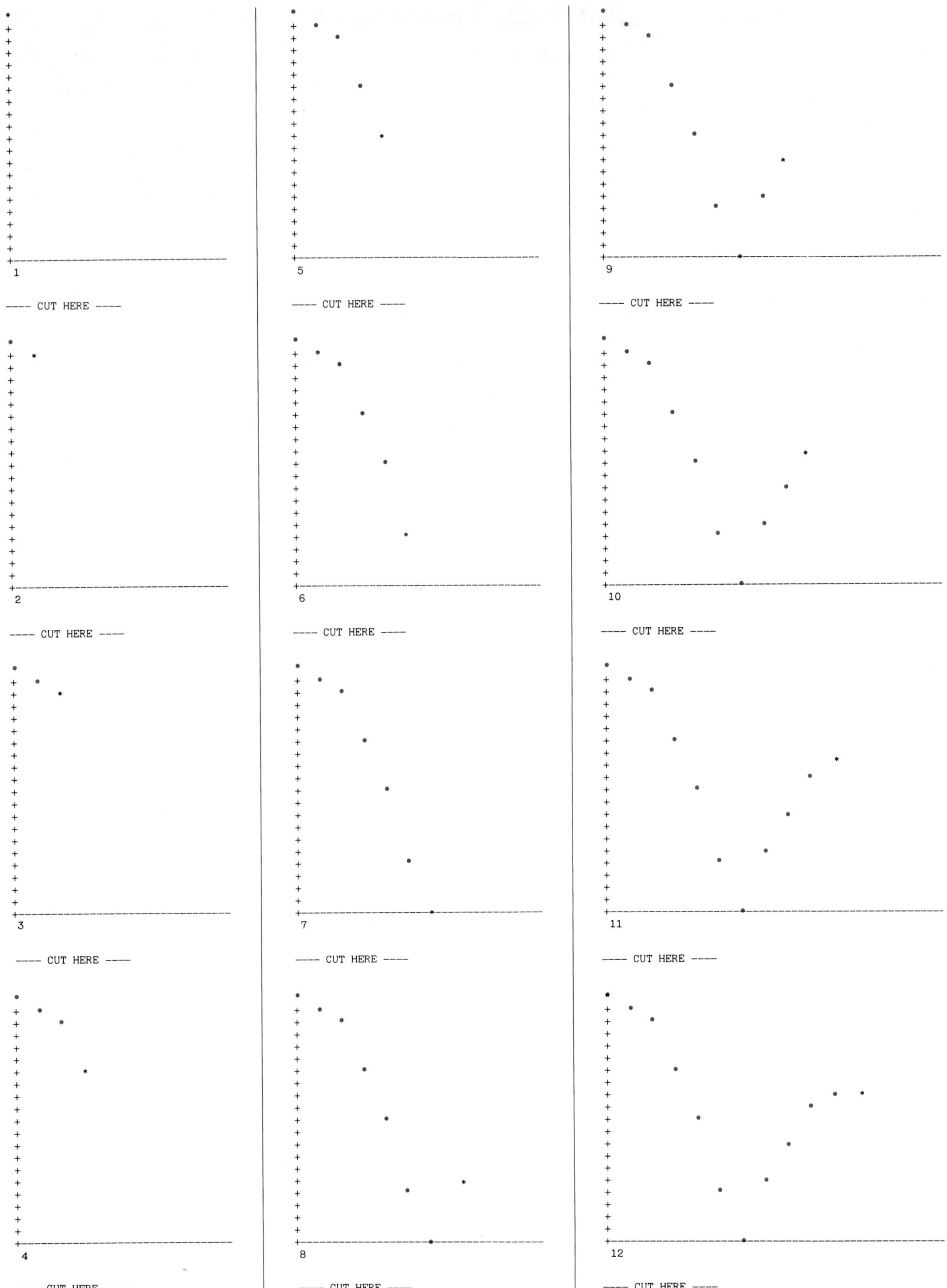
1
---- CUT HERE ----
2
---- CUT HERE ----
3
---- CUT HERE ----
4
---- CUT HERE ----
5
---- CUT HERE ----
6
---- CUT HERE ----
7
---- CUT HERE ----
8
---- CUT HERE ----
9
---- CUT HERE ----
10
---- CUT HERE ----
11
---- CUT HERE ----
12
---- CUT HERE ----

Unit 10C. The Bouncing Ball—Mathematics Needed

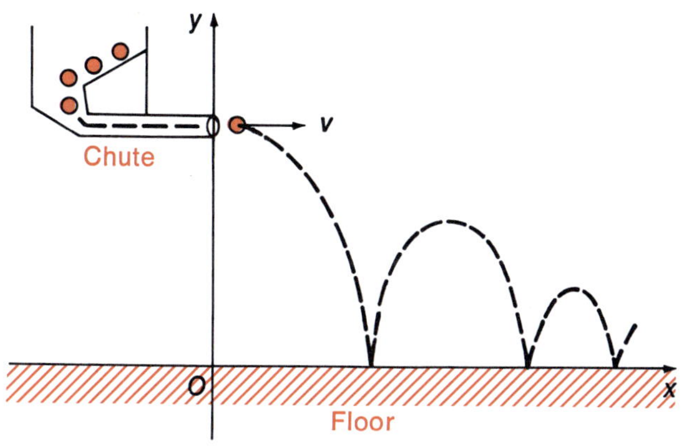

Let's examine the path of a tennis ball that is given a horizontal (parallel to the floor) velocity of v meters per second (m/s). Our picture shows one way to imagine this, with a hopper feeding the balls out of a horizontal chute. We have put a coordinate system on the picture as shown.

For our program, we'll assume that the ball leaves the chute with a horizontal velocity of 8.4 m/s. Let's examine how far the ball moves in a short time called t (think of t as the time from one snapshot to the next).

The horizontal distance moved is:

$$x_1 = vt$$
$$x_1 = 8.4t$$

The ball also moves vertically (down).

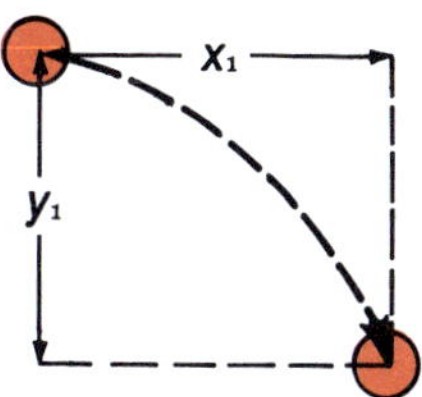

From physics we know that the vertical distance moved is approximately

$$y_1 = -\tfrac{1}{2}gt^2$$

where g is the acceleration of gravity, 9.8 m/s^2, and y_1 is negative because the acceleration of gravity is directed downward. Thus:

$$y_1 = -4.9t^2$$

If we solve the first equation for t and substitute in the second, we have:

$$y_1 = -4.9\left(\frac{x_1}{8.4}\right)^2 = -\frac{5}{72}x_1^2$$

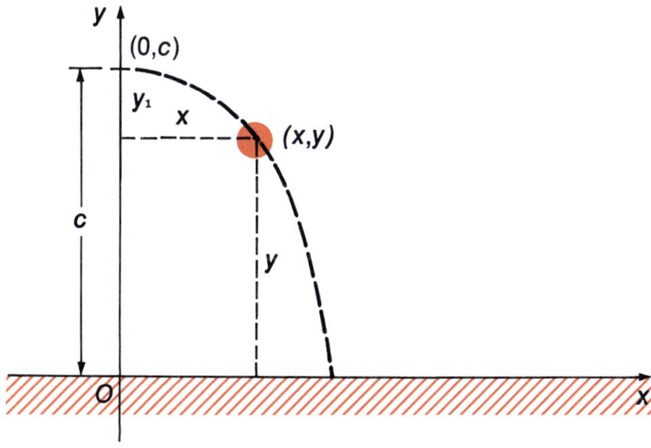

We'll have the ball start in the position $(0, c)$, where c is the ball's height above the floor. After t seconds, a snapshot would show the ball in the position (x, y), where

$$y = c - \tfrac{5}{72}x^2$$

The right side of this equation is a second-degree polynomial in x, and so a graph of y as a function of x will be a parabola.

The same argument applies to the ball's path after it bounces the first time. Its motion is still governed by the same equations except that some of the ball's vertical energy has been lost. It now follows a parabola that goes only to a height *cd*, where *d* is a number less than 1 (determined by the elasticity of the ball).

Suppose that the ball hits the floor at a point $(w, 0)$, as shown in the picture below.

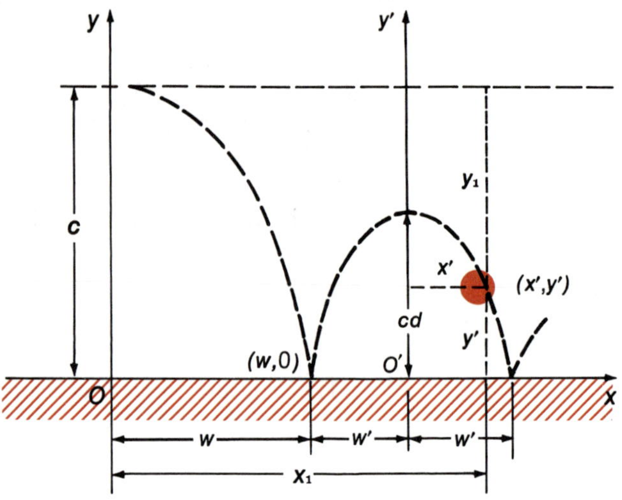

Then

$$0 = c - \tfrac{5}{72}w^2$$

or

$$w = \tfrac{6}{5}\sqrt{10c}.$$

Starting at $(w, 0)$, we compute points for the second parabola *as though* it were referred to its axis of symmetry (the y'-axis). However, x_1 is always measured from the original y-axis. All later bounces may be handled in a similar manner.

Unit 10D. The Bouncing Ball— Computer Program

10D-1: SNAPSHOT We now write a program that will produce snapshots of a bouncing ball. Since the computer prints graphs from the top down, it graphs points (x_1, y_1), where $y_1 = c - y$, instead of (x, y) as shown in the diagrams on page 159.

The snapshots on page 158 were obtained by running the following program for C = 20, S = 3, and D = 0.6. You should experiment with a variety of numbers. Save your output from this program if you want to try the animation experiment described on page 162.

```
10  PRINT "THE BALL IS DROPPED FROM WHAT HEIGHT";
20  INPUT C
30  PRINT "TYPE 2, 3, OR 4 FOR 36, 24, OR 18 FRAMES";
40  INPUT S
50  LET C1=C
60  PRINT "RATIO OF BOUNCE HEIGHT TO PREVIOUS HEIGHT (DECIMAL):";
70  INPUT D
80  PRINT
90  LET F=0
100  LET X1=0
110  LET X=0
120  LET W=6*SQR(10*C)/5
130  LET Y=C-5*X*X/72
140  LET Y=INT(Y+.5)
150  IF Y >= 0 THEN 170
160  LET Y=0
170  IF X1>71 THEN 560
180  LET F=F+1
190  LET Y1=C1-Y
200  IF X1=0 THEN 450
210  FOR Q=1 TO Y1
220  PRINT "+"
230  NEXT Q
240  IF Y1=C1 THEN 470
250  PRINT "+";TAB(X1);"*"
260  FOR Q=1 TO Y-1
270  PRINT "+"
280  NEXT Q
290  PRINT "+";
300  FOR Q=2 TO 71
310  PRINT "-";
320  NEXT Q
330  PRINT
340  PRINT F
350  PRINT "---- CUT HERE ----"
360  PRINT
370  LET X1=X1+S
380  IF X >= W THEN 410
390  LET X=X+S
400  GOTO 130
410  LET C=C*D
420  LET W=6*SQR(10*C)/5
430  LET X=-W+S
440  GOTO 130
450  PRINT "*"
460  GOTO 260
470  PRINT "+";
480  FOR Q=2 TO X1
490  PRINT "-";
500  NEXT Q
510  PRINT "*";
520  FOR Q=X1+2 TO 71
530  PRINT "-";
540  NEXT Q
550  GOTO 330
560  PRINT "FINISHED"
570  END
```

- Line 90: F will be the frame number.
- Line 120: Initial value of W
- Lines 130–140: Computes Y
- Lines 150–160: Keeps Y from dropping below the "floor"
- Line 180: Increments F
- Line 190: Computes Y1—used in graphing
- Lines 210–280: Graphs point as * except on X-axis
- Lines 290–320: Case 1. (Ball is above floor) The X-axis is printed with the symbol "-".
- Line 370: Increments X1—used in graphing
- Line 390: Increments X
- Lines 410–430: Adjusts values for next bounce
- Line 450: Prints first * at (O,C)
- Lines 470–540: Case 2. (Ball is on the floor) The X-axis is printed with the symbol "-" except for the one point where the ball is on the floor.

Unit 10E. The Bouncing Ball—Animation

Short animations can be made with flip cards or a movie camera.

To make flip-card "movies," you will need: output from program 10D-1, scissors, rubber cement, 5″ x 8″ index cards, pencil, eraser, and a large, heavy clip (optional).

1. Trim the left side of your output so that the sheet is about $7\frac{3}{4}''$ wide (so it won't hang over the 8″ card and the snapshots will be nearer the left side of the cards).

2. Put a light pencil mark on the lower left side of each index card—in the same place on each card, say 3 mm up from the bottom. This is where your "CUT HERE" line will go.

3. Spread rubber cement thinly all over one side of each index card (one card for each snapshot) and *let it dry*.
(If necessary, thin the cement out with thinner.)

4. Spread cement *thinly* all over the back of your output and *let it dry*. (Follow drawing below.)

5. Cut out your first snapshot at the "CUT HERE" line.

6. Put a piece of wax paper over your index card, covering all except the pencil mark and a little above it.

7. Looking through the wax paper, line up your trimmed snapshot with the left edge of the index card, then move it down until the "CUT HERE" line meets the pencil mark.

8. Smooth the snapshot down on the index card. Only the part not covered by the wax paper will stick.

9. Check to see if your snapshot is lined up correctly. If it is not, carefully pull it up again and try once more. If it's OK, slowly pull the wax paper out from between the snapshot and the card: pull out some, smooth down some, pull out more, smooth down, and so on.

10. Now with your finger or the eraser, rub away the rubber cement that shows around the output. It will all come off.

After you have all your snapshots cemented to index cards, stack the cards in order, clip them or grasp them firmly on the right side and riffle them with your left thumb; presto—animation!

To make movies with a camera, you will need one that has *single framing capability.* You will also need a tripod or a copy stand to hold the camera firmly. Each camera works a little differently, and so you will have to look at the instructions or get the owner to help you. You photograph each computer-generated snapshot as one frame of the movie.

For projection, splice your film into a continuous loop. This will give the illusion of a continuous supply of balls being ejected from the chute. It will also give you a chance to study the motion as long as you wish.

Try using S = 2 in program 10D-1. If the resulting motion is too fast, try making two movie frames from each snapshot.

10E-1: TRAIL (Difficult) Modify program 10D-1 so that the ball is printed as an asterisk (*), while all its previous positions are printed as periods (.), making the output look like the pictures on page 158.

10E-2: COMET (Even more difficult) Modify program 10E-1 so that only about 3 periods trail behind each asterisk, giving the ball a comet-like appearance.

10E-N, where N ≥ 3: You're on your own—have fun!

Appendix Matrix Algebra

Matrix algebra is one of the most useful branches of modern mathematics. Computers are ideal tools for working with matrices, because of the large number of computations required by matrix operations. In this Appendix we'll illustrate computer techniques for performing the most common matrix operations: addition, subtraction, multiplication, inversion, and finding integral powers of a matrix.

A *matrix* (plural, *matrices*) can be thought of as a table or array of numbers. Such tables occur naturally in both scientific and business applications. For example, suppose that the Trans-Prairie Telephone Company keeps statistics on customer preference for various telephone colors in three cities. Here's a table showing how many telephones of each color were installed in November.

NOVEMBER INSTALLATIONS

	Gotham City	Middleburgh	Prairie Junction
Black Phones	9428	1063	106
Red Phones	2346	270	46
Yellow Phones	5122	860	23
Blue Phones	1461	10	0

The numbers in this table make up a matrix. Each number is called an *entry* or an *element* of the matrix. The entry 860, which we have ringed, tells us that 860 yellow telephones were installed in Middleburgh in November. Similar meanings can be given to the other entries.

Here's how a mathematician would write the preceding data as a matrix called N:

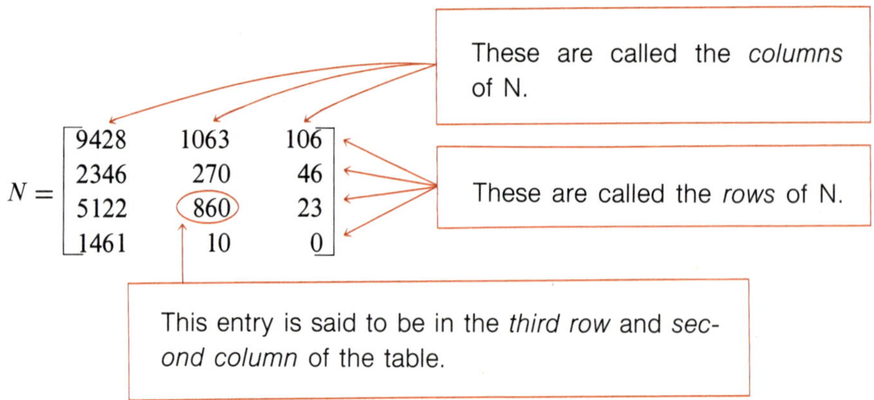

Since there are four rows and three columns, this matrix is said to have dimensions *four* by *three* (written 4×3).

Adding Matrices

DECEMBER INSTALLATIONS

$$D = \begin{bmatrix} 6328 & 2212 & 142 \\ 4123 & 20 & 0 \\ 9121 & 199 & 10 \\ 2013 & 0 & 0 \end{bmatrix}$$

If we have more than one matrix, we can define operations between these matrices. Suppose, for example, that we have stored the statistics on Trans-Prairie's December telephone installations (with the same meaning for rows and columns as in the November table) in the matrix shown at left.

We can now define the *sum* of the two matrices N and D to be another 4×3 matrix (let's call it S) such that each entry in S is the sum of the corresponding entries in N and D.

Thus:

$$N + D = S = \begin{bmatrix} 15756 & 3275 & 248 \\ 6469 & \underline{\ ?\ } & \underline{\ ?\ } \\ \underline{\ ?\ } & \underline{\ ?\ } & \underline{\ ?\ } \\ \underline{\ ?\ } & \underline{\ ?\ } & \underline{\ ?\ } \end{bmatrix}$$

EXERCISE: Calculate the missing elements of S. As you can see, S is a matrix that tells us how many telephones of each color were installed in each city in the two-month period November + December.

The most flexible approach to handling matrices in a computer is to store the value of each element in a variable with a double subscript (see page 104). The usual agreement is that the first subscript corresponds to the row, and the second subscript corresponds to the column. In our example, the variables needed for storing the data for November would be:

N(1,1)	N(1,2)	N(1,3)
N(2,1)	N(2,2)	N(2,3)
N(3,1)	N(3,2)	N(3,3)
N(4,1)	N(4,2)	N(4,3)

Similar variables would be used for the December matrix and for the Sum matrix.

A-1: MATADDI Below is a program that adds two 4 × 3 matrices. Run the program using the data from our telephone problem and verify the print-out shown on page 166.

```
10  DIM N[4,3],D[4,3],S[4,3]
20  PRINT "THIS PROGRAM ADDS TWO 4 X 3 MATRICES."
30  PRINT "INPUT THE ELEMENTS OF MATRIX N:"
40  FOR R=1 TO 4
50  FOR C=1 TO 3
60  INPUT N[R,C]
70  NEXT C
80  NEXT R
90  PRINT "NOW INPUT THE ELEMENTS OF MATRIX D:"
100  FOR R=1 TO 4
110  FOR C=1 TO 3
120  INPUT D[R,C]
130  NEXT C
140  NEXT R
150  PRINT "SUM IS:"
160  FOR R=1 TO 4
170  FOR C=1 TO 3
180  LET S[R,C]=N[R,C]+D[R,C]
190  PRINT TAB(10*(C-1));S[R,C];
200  NEXT C
210  PRINT
220  NEXT R
230  END
```

Inputs matrix N entry by entry across each row. (lines 40–80)

Inputs matrix D entry by entry across each row. (lines 100–140)

Computes entries in S. (line 180)

Prints entries in S. (line 190)

```
RUN

THIS PROGRAM ADDS TWO 4 X 3 MATRICES.
INPUT THE ELEMENTS OF MATRIX N:
?9428
?1063     First row of N
?106
?2346     Second row of N
?270
```

```
SUM IS:
 15756      3275      248
 6469       290       46
 14243      1059      33
 3474       10        0
```

A-2: MATADD2 Write a program to calculate C = A + B for matrices that can have any size up to and including 15 rows and 15 columns.

HINTS: (1) Dimension A, B, and C as 15 by 15.
(2) Allow the user to input the number of ROWS = M.
(3) Allow the user to input the number of COLUMNS = N.
(4) Change all the FOR loops to go TO M or TO N.

Notice that each portion of a program that deals with a matrix requires programming two loops—one for the rows and one for the columns. Some versions of BASIC have special MAT statements that do this work for you. These MAT statements may vary from system to system. The illustrations that are given in this book were run on the system of Time Share Corporation. Check the manual for your system, especially the MAT INPUT statement, to see if there are any differences.

MAT INPUT

Steps 40–80 of program A-1 can be replaced with:

```
40 MAT INPUT N
```

Only one question mark is printed, and you type in all the entries of N one after the other, row by row, separated by commas.
Steps 100–140 of program A-1 can be replaced with:

```
100 MAT INPUT D
```

MAT S=N+D and MAT PRINT

Steps 160–220 of program A-1 can be replaced with:

```
160 MAT S=N+D
170 MAT PRINT S
```

(Notice that you cannot compute in a MAT PRINT statement.) Here step 170 will print the columns in the five standard zones. If there are more than five columns, the results may be confusing. If a semicolon is used after S in line 170, the entries will be printed close together, but they will not be aligned in columns unless all the entries have the same number of digits.

A-3: MATADD3 This is the same as A-1 with the MAT statements substituted. Lines 170 and 180 show both print-out forms.

```
10  DIM N[4,3],D[4,3],S[4,3]
20  PRINT "THIS PROGRAM ADDS TWO 4 X 3 MATRICES."
30  PRINT "INPUT THE ELEMENTS OF MATRIX N:"
40  MAT  INPUT N
90  PRINT "NOW INPUT THE ELEMENTS OF MATRIX D:"
100  MAT  INPUT D
150  PRINT "SUM IS:"
160  MAT S=N+D
170  MAT  PRINT S
180  MAT  PRINT S;
230  END
RUN

THIS PROGRAM ADDS TWO 4 X 3 MATRICES.
INPUT THE ELEMENTS OF MATRIX N:
?9428,1063,106,2346,270,46,5122,860,23,1461,10,0
NOW INPUT THE ELEMENTS OF MATRIX D:
?6328,2212,142,4123,20,0,9121,199,10,2013,0,0
SUM IS:
 15756              3275              248

 6469               290               46

 14243              1059              33

 3474               10                0

 15756 3275 248

 6469 290 46

 14243 1059 33

 3474 10 0
```

To get a neater print-out for the matrix, you can use steps from program A-1:

```
170  FOR R=1 TO 4
180  FOR C=1 TO 3
190  PRINT TAB(10*(C-1));S[R,C];
200  NEXT C
210  PRINT
220  NEXT R
```

Subtracting Matrices

A-4 MATSUB Calculate the matrix that shows the difference in sales between December and November for each color of telephone in each city.

HINT: Change line 150 to PRINT "D – N IS:" and change line 180 in A-1 to LET S(R,C) = D(R,C) – N(R,C) or line 160 in A-3 to MAT S = D – N.

Multiplying a Matrix by a Number

To multiply a matrix by a number (called a *scalar*), multiply each entry in the matrix by that number:

$$B = \begin{bmatrix} 7 & 3 & 2 \\ 5 & -4 & -6 \\ -10 & 1 & 0 \end{bmatrix}; \qquad \tfrac{1}{2}B = \begin{bmatrix} 3.5 & 1.5 & 1 \\ 2.5 & -2 & -3 \\ -5 & -0.50 & 0 \end{bmatrix}$$

This can be done with nested FOR loops, or with the single MAT statement:

```
MAT A=(K)*B
```

Here B and A are matrices, and K is a number (or an expression). The parentheses around K are required.

A-5: MATAVG Write a program that prints a matrix showing the *average* number of telephones installed *per month* based on the November and December data.

If your version of BASIC has MAT functions, you can use BASIC statements like the following:

```
MAT S=N+D
MAT A=(.5)*S
```

These statements will give you a matrix A in which the elements are given by:

```
A(R,C)=.5*(N(R,C)+D(R,C))
```

If you don't have MAT statements, you can use nested FOR loops to write this program.

```
MAT READ
```

A-6: PERCENT1 Here's a program which calculates the percent of change from matrix M to matrix K. That is, it finds a matrix P where each entry is:

```
P(R,C)=100*(K(R,C)-M(R,C))/M(R,C)
```

$$M = \begin{bmatrix} 1 & 2 & 3 & 4 & 0 \\ 6 & 7 & 8 & 9 & 0 \\ 0 & 2 & 3 & 0 & 3 \\ 4 & 0 & 2 & 3 & 5 \end{bmatrix} \qquad K = \begin{bmatrix} 0 & 9 & 8 & 7 & 6 \\ 5 & 4 & 3 & 2 & 1 \\ 0 & 9 & 0 & 8 & 0 \\ 7 & 0 & 6 & 0 & 5 \end{bmatrix}$$

For example, the percent change from M(1, 4) to K(1, 4) is 100(7 – 4)/4% or 75%.

A difficulty arises when an entry of *M* is 0. That is, we cannot find the *percent* change from M(1, 5) to K(1, 5) because M(1, 5) = 0 and we cannot divide by zero. The program provides for this situation by assigning the "nonsense" value 9E9 (which the computer may type as 9E + 09) to the entry of *P* in such cases. This program also illustrates use of the MAT READ statement instead of MAT INPUT.

```
10  DIM K[4,5],M[4,5],D[4,5],P[4,5]
20  MAT  READ K
30  MAT  READ M
40  MAT D=K-M
50  FOR R=1 TO 4
60  FOR C=1 TO 5
70  IF M[R,C]=0 THEN 100
80  LET P[R,C]=100*D[R,C]/M[R,C]
90  GOTO 110
100  LET P[R,C]=9.E+09
110  NEXT C
120  NEXT R
130  PRINT "TABLE OF PERCENT CHANGES--K OVER M"
140  PRINT
150  MAT  PRINT P
160  DATA 0,9,8,7,6,5,4,3,2,1,0,9,0,8,0,7,0,6,0,5
170  DATA 1,2,3,4,0,6,7,8,9,0,0,2,3,0,3,4,0,2,3,5
180  END
RUN

TABLE OF PERCENT CHANGES--K OVER M

-100            350            166.667     75            9.00000E+09

-16.6667        -42.8571       -62.5       -77.7778      9.00000E+09

 9.00000E+09    350            -100        9.00000E+09   -100

 75             9.00000E+09    200         -100          0
```

> NOTES: Several matrices can be INPUT or READ in the same command: for example:
>
> MAT INPUT N,D   or   MAT READ N,D
>
> Also possible are statements such as:
>
> MAT A=A+B   or   MAT B=(K)*B

A-7: PERCENT2 Modify the program A-6 so that (1) it rounds the percentages to integral values, and (2) a percent of 0 is printed when the corresponding entries of *K* and *M* are *both* zero.

A-7X: EFFICIENT (EXTRA) Can you rewrite A-7 so that it only requires dimensioning and using three matrices? two? HINT: Could *K* also be used to hold the differences in line 40? Could *D* also hold percents in line 80?

Multiplying Matrices

The product of two matrices is defined in a way that looks strange at first, but which is quite useful. We'll discuss two programs for calculating such a product and then look at some applications of matrix multiplication.

In general, in order that there be a matrix product AB, the number of *columns* in A must be the same as the number of *rows* in B. The rule for getting the entries in the product matrix $P = AB$ is the following:

> P(R,C) is calculated as the sum of the products of the corresponding elements in the Rth row of A and the Cth column of B.

That's a pretty complicated definition, and it needs to be read carefully. To see if you understand it, check the calculations for the following example:

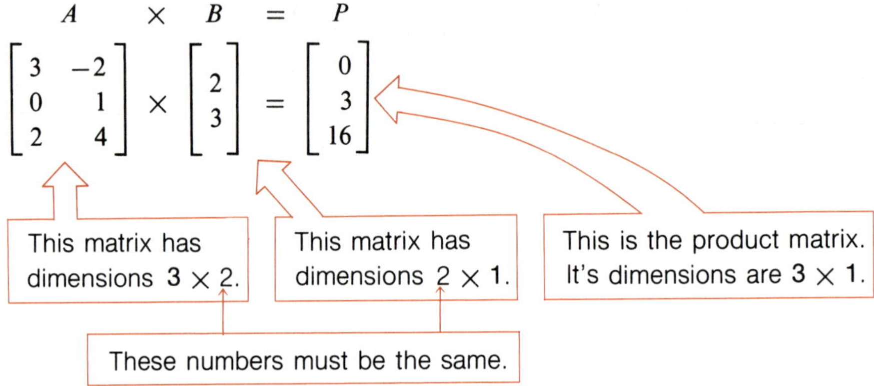

The entries in the product are found as follows:

$$P(1,1) = (3)(2) + (-2)(3) = 0$$
$$P(2,1) = (0)(2) + (1)(3) = 3$$
$$P(3,1) = (2)(2) + (4)(3) = 16$$

MAT P=A*B

A-8: MATPROD1 If your version of BASIC has MAT multiplication, the computer does all this work for you. Here's a program that illustrates the technique. As you may imagine, line 40 really involves a lot of unseen calculations. Run this program using the matrices A and B given above.

```
10  DIM A[3,2],B[2,1],P[3,1]
20  PRINT "INPUT MATRIX A, THEN B:"
30  MAT  INPUT A,B
40  MAT P=A*B
50  PRINT "PRODUCT MATRIX"
60  MAT PRINT P
70  END
```

A-9: SQMAT Modify the preceding program to multiply two 3 × 3 matrices, and then try it on matrices A and B at the top of page 171.

$$A = \begin{bmatrix} 3 & 0 & -2 \\ 4 & -7 & 1 \\ 3 & 2 & 2 \end{bmatrix} \qquad B = \begin{bmatrix} -1 & 4 & 0 \\ 8 & 2 & 1 \\ 9 & 3 & 0 \end{bmatrix}$$

Next try it on the same two matrices with the first one called B, and the second one called A. Does $AB = BA$; that is, is matrix multiplication commutative?

> NOTE: The MAT multiplication statement may *not* use the same matrix on both sides of the = sign. For example,
>
> MAT A=A*B is *not* permitted.
>
> To see why, study the next program.

A-10: MATPROD2 Write a program that multiplies two matrices without using the MAT features. Here's a partial flow chart for multiplying a matrix A of dimensions $M \times N$ by a matrix B of dimensions $K \times L$ to obtain a product P.

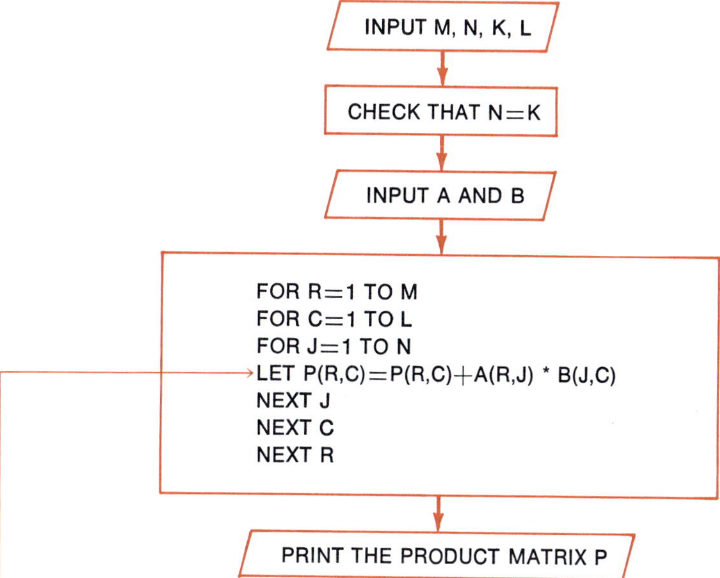

> Here's where the entries of the product matrix are calculated. To see how it works, try this algorithm on the example below, where M = 3, N = K = 3, and L = 2.
>
> $$\begin{bmatrix} a_{11} & a_{12} & a_{13} \\ a_{21} & a_{22} & a_{23} \\ a_{31} & a_{32} & a_{33} \end{bmatrix} \times \begin{bmatrix} b_{11} & b_{12} \\ b_{21} & b_{22} \\ b_{31} & b_{32} \end{bmatrix}$$
>
> $$= \begin{bmatrix} a_{11}b_{11} + a_{12}b_{21} + a_{13}b_{31} & a_{11}b_{12} + a_{12}b_{22} + a_{13}b_{32} \\ a_{21}b_{11} + a_{22}b_{21} + a_{23}b_{31} & a_{21}b_{12} + a_{22}b_{22} + a_{23}b_{32} \\ a_{31}b_{11} + a_{32}b_{21} + a_{33}b_{31} & a_{31}b_{12} + a_{32}b_{22} + a_{33}b_{32} \end{bmatrix}$$

Such a program can be used for multiplying matrices up to 10×10 without dimensioning. When you use the MAT functions, however, each matrix must be dimensioned.

Inverse Matrices

One of the most common applications of matrices is describing (and solving) linear systems. To see why, notice that we can write any linear system in matrix form as shown in the following example:

$$\begin{aligned} x_1 + x_2 + x_3 &= 4 \\ 2x_1 - x_2 + 2x_3 &= 5 \\ x_1 - 2x_2 - x_3 &= -3 \end{aligned} \quad \text{becomes} \quad \Longrightarrow \quad \begin{bmatrix} 1 & 1 & 1 \\ 2 & -1 & 2 \\ 1 & -2 & -1 \end{bmatrix} \times \begin{bmatrix} x_1 \\ x_2 \\ x_3 \end{bmatrix} = \begin{bmatrix} 4 \\ 5 \\ -3 \end{bmatrix}$$

Now if we let

$$A = \begin{bmatrix} 1 & 1 & 1 \\ 2 & -1 & 2 \\ 1 & -2 & -1 \end{bmatrix}, \quad X = \begin{bmatrix} x_1 \\ x_2 \\ x_3 \end{bmatrix}, \quad \text{and } B = \begin{bmatrix} 4 \\ 5 \\ -3 \end{bmatrix},$$

then this linear system can be written as:

$$AX = B$$

Thus,

$$X = A^{-1}B,$$

where A^{-1} is the inverse of A, that is, where $A^{-1}A = I$ and $IX = X$.

MAT V=INV(A)

Finding A^{-1} is a bit complicated, and we won't explain it here. However, some versions of BASIC do matrix inversion automatically.

A-11: VERIFY This program will find the inverse of A and then the products AA^{-1} and $A^{-1}A$ when A is a 3×3 matrix.

```
10  DIM A(3,3),V(3,3),P(3,3),Q(3,3)
20  PRINT "INPUT MATRIX A:"
30  MAT INPUT A
40  MAT V=INV(A)
50  MAT P=A*V
60  MAT Q=V*A
70  MAT PRINT P,Q
80  END
RUN

INPUT MATRIX A:
?1,1,1,2,-1,2,1,-2,-1
 1              5.96046E-08      0

 2.38419E-07    1                0

 1.19209E-07    0                1

 1.             0                1.19209E-07

 0              1                0

 0              0                1
```

Notice that both product matrices are very close to the identity matrix:

$$I = \begin{bmatrix} 1 & 0 & 0 \\ 0 & 1 & 0 \\ 0 & 0 & 1 \end{bmatrix}$$

A-12: SOLVE1 The following program will find the solution of a system of three linear equations in three variables, expressed in matrices as

$$X = A^{-1}B$$

where A is nonsingular (which means that A has an inverse).

```
10  DIM A(3,3),B(3),X(3),V(3,3)
20  PRINT "INPUT MATRICES A,B:"
30  MAT INPUT A,B
40  MAT V=INV(A)
50  MAT X=V*B
60  PRINT "SOLUTION SET:"
70  MAT PRINT X
80  END
RUN

INPUT MATRICES A,B:
?1,1,1,2,-1,2,1,-2,-1,4,5,-3
SOLUTION SET:
 1.

 1

 2
```

Notice in line 10 that although B and X are 3×1 matrices, they may be dimensioned simply as B(3) and X(3) (such matrices are also called *vectors*).

A-13: SOLVE2 Change line 10 of program A-12 to allow for 4 equations in 4 variables, and solve the system given for program 6C-3, (page 112).

NOTE: Programs A-12 and A-13 solve linear systems of exactly the same kind as discussed in programs 6C-2 and 6C-3. They look simpler only because MAT INV does a *lot* of work unseen by the programmer. In fact, inversion of an N $\times$ N matrix is usually done by a procedure that essentially uses the Gaussian elimination method explained in section 6, but applied to the N columns of the identity matrix. If you'd like to compare the accuracy of A-12 and 6C-2, here's a good test problem:

$$\begin{aligned} x + .50y + .33z &= 31.80 \\ .50x + .33y + .25z &= 21.96 \\ .33x + .25y + .20z &= 16.98 \end{aligned}$$

The exact solution is (6, 12, 60).

Communication Matrices

The diagram below depicts the "safe" channels of communication between seven members of an international ring of special agents. The arrows on the lines mean that messages are authorized only in the directions marked.

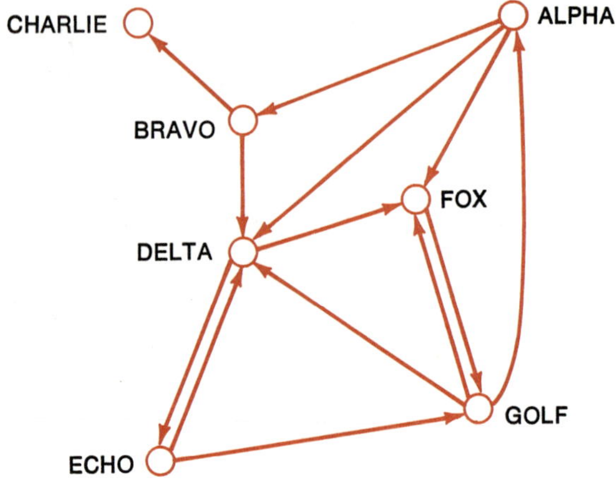

A diagram of this sort is called a *directed graph,* or *digraph.* A line joining two points (or *nodes*) in the digraph is called a *branch.* As can be seen, there are many possible paths over which information can flow in such a digraph, and some paths have more branches than others. We'd like to use the computer to help us find the number of possible paths consisting of one branch, the number with two branches, the number with three branches, and so on, for any *pair* of agents.

The "security" of any path can be assumed to become poorer the more branches there are. We'd therefore also like to study how the number of paths depends on the number of branches allowed in a path, and then use the computer to find out which agents cannot communicate if we limit paths to some "safe" number of branches.

Our first problem is to describe our "picture" (the digraph) to the computer. We do this by defining a matrix M as follows:

	A	B	C	D	E	F	G
A	0	1	0	1	0	1	0
B	0	0	1	1	0	0	0
C	0	0	0	0	0	0	0
D	0	0	0	0	1	1	0
E	0	0	0	1	0	0	1
F	(0)	0	0	0	0	0	1
G	(1)	0	0	1	0	1	0

$= M$

This says that there is one single-branch path over which agent Golf can communicate information to agent Alpha.

This says that there are no single-branch paths over which agent Fox can communicate information to agent Alpha.

In general, M(I,J) = 1 if agent I can communicate with agent J over a single-branch path; otherwise it's 0.

Since M is a square matrix, it can be multiplied by itself. We write:

$$M \times M = M^2$$

Now a remarkable property (not proved here) of M^2 is that it will show which agents can communicate with others by paths that have two branches (such a path might be called a two-step communication). Not only that, M^2 will tell us how many such two-step paths there are.

A-14: SPY Write a program to calculate M^2, and use the result to find all the two-step communication paths.

Here's part of the solution:

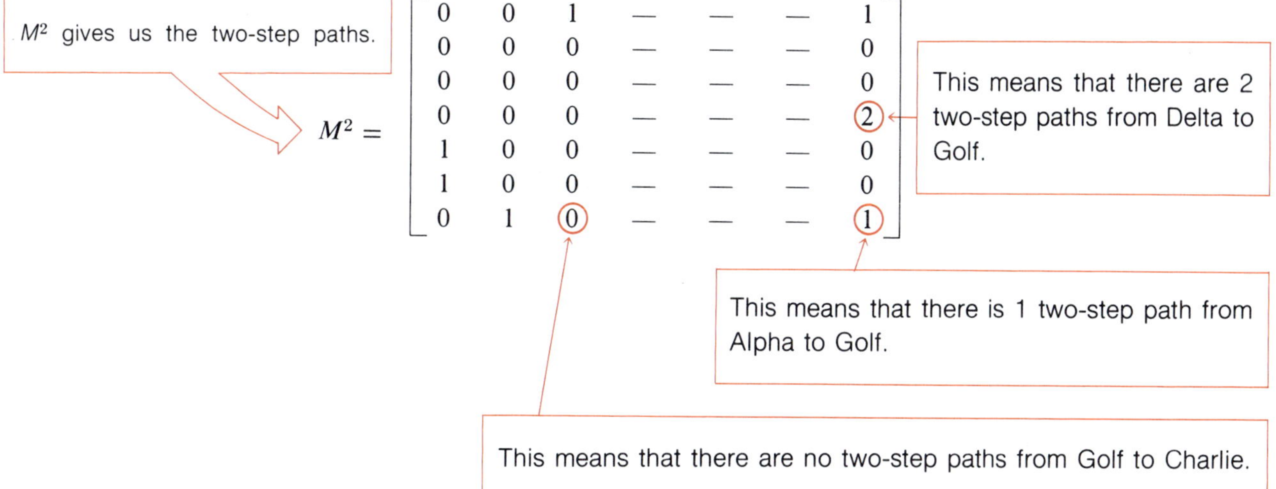

After you have calculated the rest of M^2, you should look at the digraph, and check that you can find all the two-step paths predicted by M^2.

A-15: SUPERSPY Calculate the number of three-step and four-step paths between agents.

HINT: Calculate $M^3 = M^2 \times M$ and $M^4 = M^3 \times M$.

A-16: LONELYSPIES Which agents cannot possibly communicate with each other if we only permit paths with (1) one step? (2) two steps or less? (3) three or less? (4) four or less?

HINTS: (1) Look for the zeros in M.

(2) Look for the zeros in $M + M^2$.

(3) Look for the zeros in $M + M^2 + M^3$.

(4) Look for the zeros in $M + M^2 + M^3 + M^4$.

Some Additional Ideas

(1) It would be interesting to list (in order) the pair of agents having the largest number of backup communication paths available to them, the pair with the next largest etc. This gets much more interesting with a large network (say 20 agents).

(2) Suppose that one agent in a large network is put out of business. This means that some paths disappear from the resulting digraph. Try taking higher powers of the new *M* to see which new paths can be established without hiring a new agent.

(3) The preceding ideas also apply to the transmission of communicable diseases and to airline or freighter scheduling. Research one of these areas, and develop some computer programs to analyze the situation.

(4) Long distance telephone calls take place over a network that can be described by a directed graph. When one of the branches in this network is busy (all the lines in use), an automatic switching computer tries to find an alternate route before giving a busy signal. Write a program that simulates this process for a network of 20 cities. Input the city from which calls are being made, and the destination city (for example, New York to Chicago). Output the path selected (for example, New York to Philadelphia to Cleveland to Chicago). Have the program loop on this process until a message like "No more paths from New York to Chicago are possible" is printed.

When you first write the program, assume that each branch in the network can handle only one call. Later you can try assigning "capacities" to each branch. For example, assigning a capacity of 10 to the New York-Philadelphia branch means that 10 calls that originate in New York can simultaneously use this branch before a new branch is needed. An even more complicated simulation would be one that assigns random "times" to each call. Incidentally, if you find that a call from New York to Chicago goes by way of Texas, don't be surprised. This is the way it really happens in practice.

index

CDEFGHIJ–SM–7987